THE MANITOBA RECORD SOCIETY

THE MANITOBA RECORD SOCIETY PUBLICATIONS

VOL **X** UME

GENERAL EDITOR
J.M. Bumsted

ASSOCIATE EDITOR
H.J. Mays

The Manitoba Record Society
Winnipeg
1991

The

MODERN BEGINNINGS OF SUBARCTIC ORNITHOLOGY: CORRESPONDENCE TO THE SMITHSONIAN INSTITUTION

1856 - 1868

EDITED AND INTRODUCED BY
Debra Lindsay

Canadian Cataloguing in Publication Data

The modern beginnings of subarctic ornithology
(Manitoba Record Society publications; v.10)

ISBN 0-9692101-3-2

1. Ornithology - Canada, Northern. 2. Naturalists - Canada - Correspondence. 3. Fur traders - Canada - Correspondence. 4. Smithsonian Institution - Natural history collections. 5. Rupert's Land - Description and travel. 6. Rupert's Land - Social life and customs. 7. Natural history - Canada - History - 19th century. I. Lindsay, Debra J. (Debra Jean), 1955 - II. Manitoba Record Society. III. series.

QL685.5.N67M63 1991 598.29719 C92-098009-0

CONTENTS

**To my parents,
Ann and Gordon Lindsay**

A back view of the Smithsonian Institution, circa 1860. The "Castle" was the residence of the Henry family and the repository for all the specimens and data sent Assistant Secretary Baird and Secretary Henry. Smithsonian Institution Photo #9748A.

Robert Kennicott (1835–66) in northern garb. Kennicott posed for this photograph after returning home from the Mackenzie River District. He chose to be portrayed as a "frontiersman," whereas his northern confreres preferred to be photographed while dressed in their most "civilized" attire -- the ties, collars and jackets befitting their positions as gentlemen. Smithsonian Institution Photo #43604.

INTRODUCTION

In 1859, Robert Kennicott (1835-66), a young naturalist from Illinois, went north into the Mackenzie River District of the Hudson's Bay Company's territory in search of scientific specimens for the Smithsonian Institution of Washington, D.C. Because Kennicott was instructed to collect the birds, eggs, mammals, linguistic, and ethnographic material needed to provide the empirical foundation for what was hoped would become the most comprehensive research facility to be established in North America, his expedition differed from most earlier expeditions into Rupert's Land. Scientists and collectors had been permitted to collect specimens and data in the Hudson's Bay Company territories prior to Kennicott's visit, but science and exploration were typically subordinated to political or economic interests. Those few expeditions that were more purely scientific, produced significantly smaller collections than those made on behalf of the Smithsonian Institution. Collections made by David Douglas, a Scottish botanist who travelled in the northwestern United States and southern Rupert's Land in the 1820s and 1830s, as well as collections made in the 1840s by Karl Andreas Geyer, the editor of the *London Journal of Botany*, and collections made by Joseph Burke, a gardener from the Royal Botanical Gardens at Kew, were dwarfed by those received at the Smithsonian.[1] Between 1859 and 1868, the Smithsonian Institution received more than 12,000 natural history specimens, several volumes of field notes, and some ethnographic accounts from Rupert's Land. Moreover, northern Indian and Inuit artifacts constituted some of the earliest useful anthropological specimens deposited at the Smithsonian, while northern ethnographies were amongst the first descriptions of "exotic" cultures committed to paper for reasons other than the obviously practical ones advanced by missionaries, colonial administrators, or fur traders.[2] Northern zoological specimens were equally important. The skins, skeletons, and embryos of birds and mammals, as well as the fish, reptiles, amphibians, insects, shells, and eggs sent south, verified the data contained in contemporary treatises on northern natural history, as well as providing further material for research on the distribution, migration, and breeding habits of North American fauna. Studies of speciation, demographics, and systematics were facilitated by northern zoological specimens, as was microtaxonomy or the identification of species, genera, and families. Specimens collected in the Hudson's Bay Company (HBC) territories provided unsurpassed and much needed data for the natural and the social sciences.

THE HISTORICAL CONTEXT: SCIENCE IN RUPERT'S LAND BEFORE 1859

Between 1860 and 1880, the American taxonomists who were identifying and classifying North American fauna, had unprecedented access to northern specimens. Natural history specimens had been sent to Europe from the HBC's North American territories during the eighteenth century, but the Company was trying to develop a trans-Atlantic trade in furs for

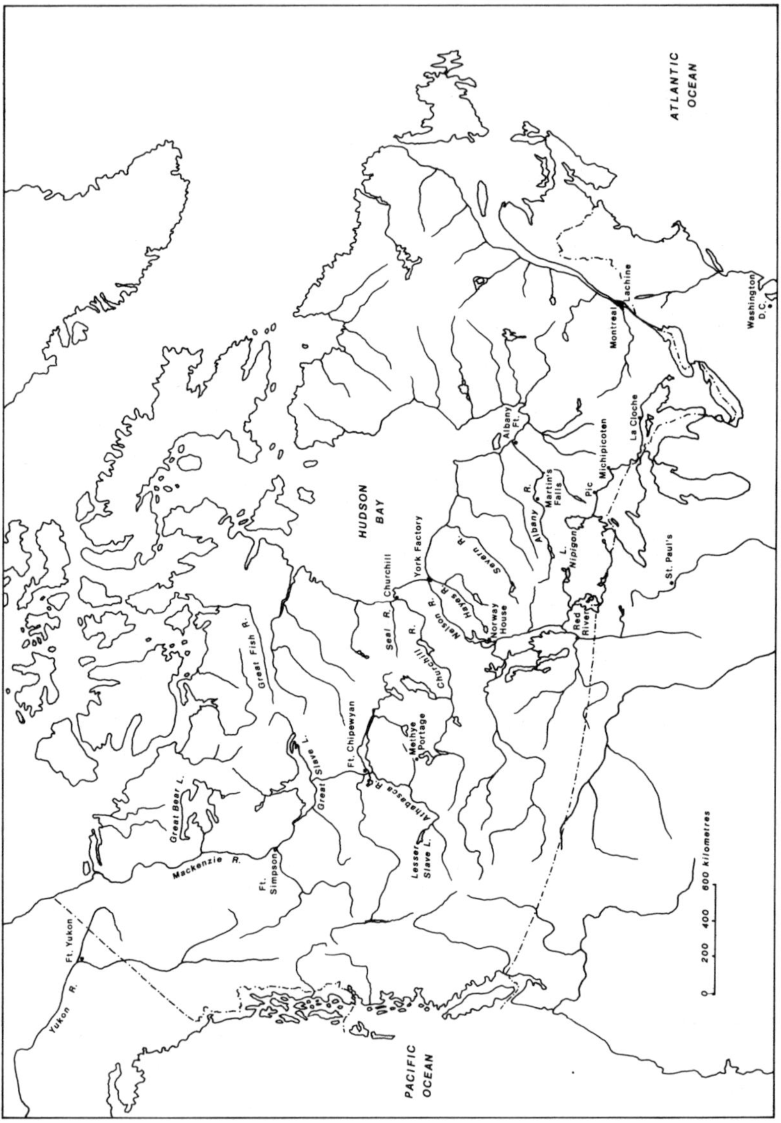

Rupert's Land and British North America before Confederation.

much of its first one hundred and fifty years of existence and science was only supported if its utility was obvious and immediate. Although five of the eighteen founding members of the HBC, as well as several company shareholders, were Fellows of the Royal Society, support for scientific pursuits was minimal for many years.[3] Corporate policies were pragmatic. Although support for a venture that combined resource exploitation, merchant capitalism, and territorial imperialism reflected the aspirations of individuals who accepted the benefits of the Scientific Revolution and welcomed the commercial opportunities offered by the Age of Discovery, scientific activities were always subsidiary to legal and commercial considerations. Profit was the guiding principle; the Company concentrated on the discovery, consolidation, and expansion of North American trading territories.

Because the Company's presence in North America was precarious until after 1675, when full-time replaced seasonal occupation at the Bay, unofficial or non-trade related uses of employees or resources was seldom permitted. Despite requests for aid, the Royal Society received slight assistance in its attempts to prosecute its "philosophical observations."[4] Moreover, when the HBC established itself in North America, the transfer of traditional European rivalries to Hudson's Bay made support for "pure" science even less attractive. Ironically, however, it was the French-English disputes which induced the Company to venture inland in search of furs. The HBC sponsored an expedition under Henry Kelsey after the French seized their trading posts on the Moose and Albany Rivers in 1686. While travelling inland, via the Winnipeg and Saskatchewan water systems, Kelsey obtained much data on the prairies and its native inhabitants that was of scientific-commercial significance.

Exploration had few attractions as long as the trade was safe from interlopers, but the conservatism of the Company was also indicative of a corporate reaction to the financial and human losses incurred when scientific ventures were undertaken. Between £7000 and £8000 of Company funds had been lost when an expedition under Captain James Knight went missing while searching for the North West Passage in 1719.[5] This episode reminded the Company of the hazards of exploration, and "... led the Committee to frown upon exploratory ventures unless they were closely and obviously connected with the maintenance and the enlargement of trade."[6] Aside from one unsuccessful voyage in 1772, and the annual voyages made by Company ships between London and the Bay, naval exploration was abandoned. Searches for a northern passage were eventually resumed in 1769, but a mandate was given for overland rather than naval exploration, and Samuel Hearne set out in search of the northwest passage nearly fifty years after Knight's tragic accident.

Experience with ice-bound seas deterred the HBC from risking either men or money on exploration in northern waters, but others were less reticient. Arthur Dobbs, the Engineer-in-Chief and Surveyor-General of Ireland, was one such individual.[7] He blamed the Company for Britain's failure to find the North West Passage, and he contended that the North West Passage not only existed, but that the discovery of the passage had thus far been purposefully prevented by the "Monopoly and Avarice of [the] Hudson's Bay Company."[8] Dobbs overestimated the ease of navigability through northern waters, but by 1744 he was so convinced that there was an elaborate plot to keep secret all information about North America and to prevent access to Rupert's Land, that he sold subscriptions to finance naval expeditions in 1746 and 1747.

These expeditions did not find the Passage, and Dobbs could not claim the £20,000 reward offered by the British government in 1745 to the discoverer of the North West Passage. He therefore regrouped and refocussed his attacks on the HBC. In so doing he brought to the forefront Rupert's Land's earliest natural history collectors. When Dobbs petitioned the King for a land grant analogous to that given the HBC, he provoked a parliamentary enquiry into the validity of the HBC Charter which made public, for the first time, Company activities in North America.[9] Chief Factor James Isham and another HBC employee, Alexander Light, returned to England to testify on behalf of the HBC. Journals kept by Isham were used to refute Dobbs' claim about the feasibility of settlement in, and navigability through, Rupert's Land. Isham's journals also contained important ethnographic and zoological information, while specimens collected by Isham and Light were used by the European naturalist, George Edwards, in the compilation of *A Natural History of Uncommon Birds* (1743-51).[10]

Within a decade, exploration, science, and commerce intersected again when the HBC decided to expand inland. From the late 1750s until 1821, HBC exploration was directed largely towards opening the Northwest to the fur trade, and to outmanoeuvering Canadian and American competition. During this period, Company officials agreed to allow their overseas employees to collect natural history specimens on behalf of the Royal Society, and in the early 1770s HBC traders collected geographical, ethnological, zoological, and botanical data as they moved inland from the Bay. John Reinhold Forster, the naturalist noted for his appointment to the Cook expedition as the first paid scientist specifically charged with representing British imperial interests, examined their specimens for his *Catalogue of the Animals of North America* (1771).[11] Thomas Pennant, who was a personal friend of the Vice-President of the Royal Society and the author of four important books on zoology, including *Arctic Zoology* (1784-85), also examined the specimens from Rupert's Land.[12] Moreover, a "Committee on Natural History" that was struck in 1772 to deal with the HBC specimens became a standing committee of the Royal Society in 1773. This committee was responsible for describing and distributing the natural history specimens coming out of Rupert's Land, and it was chaired by Samuel Wegg, a future governor of the HBC.[13]

By the end of the eighteenth century, despite a certain wariness of science and exploration, the HBC deserved some credit for having promoted the sciences.[14] Even though the London Committee of the HBC was more concerned with commerce and economics than with science, concessions had been made. Zoological and anthropological reports, as well as specimens, had been filtering into the Society since 1681, and the Society had gained access to logs kept on Company ships, as well as interviewing Company captains in order to obtain navigational information. Society scientists had received data on meteorological and magnetic phenomena that had been kept by Captain Christopher Middleton while he was employed on Company ships between 1721 and 1729, and the Company had also permitted William Wales, mathematician and astronomer, to stay at Fort Prince of Wales in order to document the parallax of the transit of Venus (3 June 1769) for the Royal Society. In 1782, Samuel Wegg, a long-time Fellow of the Royal Society, became Governor of the HBC, and he made Company records available to cartographers and geographers such as Aaron Arrowsmith and Alexander Dalrymple.[15] He also opened Bay waters to British exploration. Under his auspices, the London Committee permitted publication of the journals kept by Samuel Hearne between 1769 and 1772 while he was searching for copper deposits west of Churchill. Indeed, Wegg and his colleagues, Thomas Pennant and William Wales, were ultimately responsible for the publication of these journals. Hearne only agreed to revise them at their urging.[16]

Wegg's governorship also witnessed a subtle but significant shift in the foundation of HBC-sponsored science. During the 1780s, the Montreal-based North West Company (NWC) expanded aggressively across the northwest, forcing the HBC to do the same; by 1820 posts dotted the northwest from the bottom of the Bay and the Great Lakes to the Pacific Coast and the Athabasca. During this period the Company finally recognized that the data and data-gathering techniques used by scientists might also be useful for commercial purposes. In the struggle to out-trade Canadian competitors the London Committee became aware of the necessity of comprehensive and accurate information on all facets of the overseas trade, and in 1814 it sent instructions intended to improve the transmission and the quality of information reaching London.[17] Descriptions of the geography, the topography, and the climate of each region, as well as maps noting the navigability of the rivers were to be sent to the London Committee. The type and quantity of natural produce growing in each district were to be described, along with discussions of trading returns. Information about the posts within each district, including an assessment of the potential for crop cultivation, and descriptions of the general condition and locations of the buildings was wanted. The Committee was also interested in information on the Indians. Chiefs and hunters were to be identified individually. Information on how many families or hunters frequented HBC and NWC posts, the locations of hunting areas, and if Indians had notions about private property was also desired. Similar data had previously been made available to European scientists through the philanthropy of the Company and its overseas officers, but by the early nineteenth century Company officials had come to appreciate the utility of science and exploration. This new attitude made science and exploration more subservient than ever to corporate objectives, however.

An analogous ethos also prevailed within the British government. Science and exploration were harnessed to British imperialist goals. When the £20,000 reward first offered in 1745 was reinstated in 1818, exploration in search of a passage through North America to the

Orient escalated.[18] In that very year the British navy sent John Ross and William Edward Parry on their first voyages to the Arctic, and their two-ship expedition marked the beginning of the second phase in the search for the northern passage. Within six months of their return to England, the British Admiralty dispatched another northern naval expedition and added land exploration to naval searches for the North West Passage.

In 1819, Captain John Franklin was sent to survey the northern coastline by land. Parry commanded two more naval expeditions in 1821-23 and 1824-25, and during the latter expedition Franklin was sent on a second overland expedition to survey the northern coast. Captain Frederick William Beechey simultaneously guided his ship around Cape Horn up to Alaska. Beechey was supposed to reconnoitre with Franklin and Parry along the northern shore, but this did not happen. Franklin and his crew did not arrive at the predetermined site, but they did return home intact after spending three years in the north. Such was not the case in 1845, however, when Franklin commanded a naval expedition in search of the North West Passage. His ships were last seen less than three months after leaving London on 19 May, 1845. Numerous expeditions were consequently dispatched to find the lost Franklin expedition, and the Franklin searches provided as potent an incentive for Arctic exploration as had the elusive North West Passage.[19] The British sent sixty-eight expeditions with the express purpose of finding Franklin. Even the HBC and private citizens assisted in the search. Between 1850 and 1870, the HBC sponsored three Franklin search expeditions and directed all of its officers to assist British searchers.[20] Henry Grinnell of the United States also sponsored two Franklin searches (1850-51, 1853-55) in response to an appeal from Lady Franklin. The first Grinnell expedition was commanded by Captain Edwin Jesse De Haven, and the second was led by Elisha Kent Kane.

Two of the HBC-sponsored searches were commanded by Dr. John Rae (1813-93), a relative newcomer to Arctic explorations. He had gone on his first northern expedition soon after Franklin went missing. Rae's first expedition was part of the HBC's belated efforts in search of the North Passage, and a continuation of the work begun by Peter Warren Dease and Thomas Simpson in 1837. Dease and Simpson were ordered to survey the northern shore of North America and to search for a passage through northern waters to the Pacific. When Simpson died in the spring of 1840, Rae was chosen as his replacement.[21] For his 1846 expedition Rae was ordered to survey the northern shore while searching for a passage to the Pacific; he was instructed to "...complete the geography of the northern shore of America," and he received specific directions to:

> ... attend to botany and geology; to zoology in all its departments; to the temperature both of the air and of the water; to the conditions of the atmosphere and the state of the ice; to winds and currents; to the soundings as well with respect to bottom as with respect to depth; to the magnetic dip and the variation of the compass; to the aurora borealis and the refraction of light. You will also, to the best of your opportunities, observe the ethnographical peculiarities of the Esquimaux of the country[22]

Acting on instructions such as these, northern explorers uncovered much about Rupert's Land.

Although relentlessness rather than success tended to characterize searches for the North West Passage, two decades of doggedness had produced much geographical data on northern North America. British expeditions also provided the data needed for the preparation of several scientific accounts on the zoology and botany of North America. Appendices consisting of "Zoological" and "Geological Memoranda," a "Botanical Appendix," and data on magnetism and meteorology, were attached to the published account of John Ross's first voyage.[23] Information on the physical sciences accumulated during the three years that Franklin's crew spent in the north filled almost one hundred and fifty pages of text attached to the travel narrative of Franklin's second expedition (1828); information collected on the natural sciences warranted a separate publication.[24] *Fauna Boreali-Americana* contained descriptions of northern mammals, birds, and fishes, while *Flora Boreali-Americana* contained descriptions of the botanical specimens collected by the Franklin Expedition.[25] These monographs remained the standard works on northern natural history for several years, even though neither botanical nor zoological collections had been the primary motivation underlying Franklin's explorations. Despite the wealth of scientific information produced by the Franklin expeditions, science continued to be the by-product of imperial and commercial dictates.

Explorations undertaken in Rupert's Land by John Henry Lefroy (1817-90), Henry Youle Hind (1823-1908) and Captain John Palliser (1807-87) also coupled science with political and commercial imperatives. Lefroy and his assistant, William Henry, spent 1843 and 1844 searching for the polar or "magnetic north" on behalf of the magnetic observatory built in Toronto by the British Royal Engineers.[26] Data on meteorological and magnetic phenomena was requisite to improvements in the navigational sciences, but Lefroy's expedition was also an early manifestation of the obsession to find the North Pole even though polar searches only superseded the quest for the North West Passage after 1860. Similarly, expeditions sent to the prairie west in 1857 were supposed to ascertain the "true" characteristics and agricultural feasibility of a region which might become part of the British Empire and the Canadian state. When the licence for Exclusive Trade granted the HBC in 1821 came under Parliamentary scrutiny in 1857, company critics and Canadian expansionists were ill-prepared to present their case before the British government and the company's rights in Rupert's Land were confirmed. Not one Canadian-sponsored expedition had ventured beyond Lake Superior.[27] The Canadian government consequently sent George Gladman and Henry Youle Hind to obtain data on the northwest that was untainted by the HBC. The British sponsored a similar expedition under Captain John Palliser.

Despite all of this exploratory activity, the collection of northern natural history specimens undertaken in 1859, when the Smithsonian Institution sent Robert Kennicott to the Mackenzie River District, was the first large-scale, apolitical, and noncommercial scientific study of Rupert's Land.[28] The utilitarian or humanitarian motivations that had characterized previous expeditions into Rupert's Land did not prompt Smithsonian scientists to gather data on the north. They were not concerned with territorial expansion, transportation routes, resource exploitation, or the suitability of the north for settlement. Smithsonian scientists wanted northern specimens for research, for enumeration, and for display.

Kennicott's expedition was empirically based, and in many ways, that empiricism anticipated the nature of late nineteenth century exploration and field studies in the far north. But Kennicott's expedition also anticipated the increasingly American origins of northern exploration. Excluding countless expeditions sent to survey and explore Alaska after 1867, the United States sent thirty-four scientific expeditions into the north between the first Smithsonian sponsored explorations in 1859, and 1909, when Peary fixed the location of the magnetic pole.[29] Moreover, several American expeditions went to British North America (the North West Territories) independent of the Polar Expeditions. The American Museum of Natural History, the University of Iowa, the United States Department of Agriculture and the Carnegie Museum sponsored northern biological and anthropological expeditions. Although boundary disputes focussed Canadian attention on the north in the last quarter of the nineteenth century, exploration of more southerly regions in western Canada claimed most of the meagre budget and the trained personnel of the publicly-funded Geological Survey. Between 1887 and 1910, the Dominion government of Canada and the Geological Survey sponsored only nineteen expeditions to survey or patrol their newly acquired possessions.[30] Scientific expeditions were also fielded by France (two), Germany (four), Norway (three) and Britain (six), but taken together with Canadian sponsored research, they just matched the number of expeditions originating out of the United States. Kennicott's explorations in Rupert's Land are therefore important not only because they mark the beginning of a very productive period in the history of science in the north, but because they exemplify the American mandate and the continentalist orientation of northern exploration in the second half of the nineteenth century.

THE LETTERS

While the field notes and specimens sent south from the Mackenzie River District were reckoned to be some of the most important contributions received by the Smithsonian Institution during the Civil War years, the correspondence between northern collectors and Smithsonian scientists has not received much attention. Scholarly interests are primarily responsible for such an oversight, but to the extent that inattention is due to inaccessibility, this volume will make available approximately sixty per cent of the letters sent south from Rupert's Land.

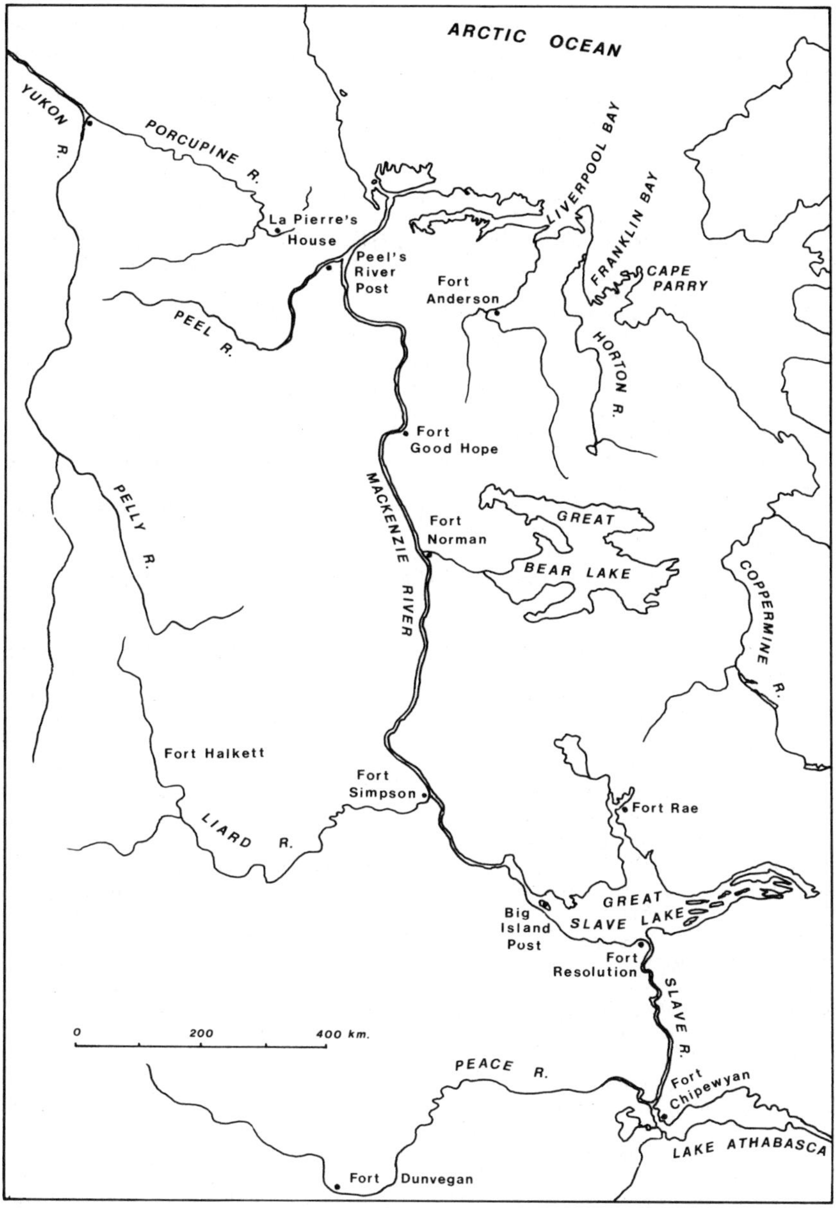

The Mackenzie River District

Because these letters were private communications rather than official business, they are forthright, chatty, and useful to a variety of scholars. These letters not only provide a fairly continuous and full account of one of the first attempts to compile scientific data on northern North America, they contain information of interest to labour historians, social historians, and native historians. For example, daily life in a northern nineteenth century outpost of the Hudson's Bay Company is described, as are the interpersonal relationships between fur traders, missionaries, native peoples, and outsiders such as Robert Kennicott. Evidence of the utilization of the local labour hierarchy for scientific activities can also be found in these letters. Reference to the role that native collectors played in the accumulation, preparation, and identification of specimens is scattered throughout the letters received in Washington between 1856 and 1868.

The wealth of information pertaining to the history of North American science "in the field" is, however, without doubt the most striking feature of these letters. They contain much of interest to students of zoological taxonomy and systematics. Descriptions of habitats and habits, as well as examples of lay usage of nineteenth century scientific nomenclature are dispersed throughout the letters. Written during the formative period in the history of the social sciences, the correspondence contains information relevant to the theoretical and methodological developments in physical anthropology, evolutionary theory, historical linguistics, and museum anthropology. Although considerable amounts of descriptive and linguistic data were submitted by northern fieldworkers, even more information on northern native peoples was transmitted through their correspondence with the Smithsonian.

Also exposed in these letters are the biases, prejudices, and world views of these early ethnographers and collectors. Unlike twentieth century anthropologists who have been, and are, conscious of the need for sensitivity towards their subjects, the attitudes that HBC fur traders and even trained fieldworkers such as Kennicott held regarding Indians and Inuit were rarely so informed. Although the labour and knowledge of indigenous peoples was irreplaceable in the collecting, preserving, packing, and transporting of natural history specimens, the dependency between observer and observed seldom manifested itself in positive depictions of northern native peoples. Northern correspondents instead described the trials and tribulations of "doing" fieldwork, and most often described the "problems" associated with native collectors in their letters to the Smithsonian Institution. Ironically, it is through these criticisms, which were so regular and gratuitous as to form a virtual convention of condemnation, that the large role played by native peoples in the acquisition of zoological and anthropological specimens emerges most decisively. Anecdotal evidence of native scientific activities is also, however, supported by more quantifiable data. Comments about the costs of specimens and local labour confirm the *ad hominem* remarks attesting to native involvement.

Many other factors also merited comment by northern correspondents: long days spent in the field, cold winters, hordes of mosquitoes, and isolation from libraries, laboratories, and like-minded individuals. But these letters not only describe the sacrifices people were willing to make on behalf of science, they also provide a glimpse into its attractions. More specifically, they underscore the lure of membership in an elite social group. The Smithsonian Institution, like the philosophical and scientific societies that proliferated during the eighteenth and nineteenth centuries in North America, performed a variety of social functions.[31] It provided a formalized setting for scientific activities; it served as a vehicle for professional advancement as well as for entertainment; it facilitated the exchange of ideas and specimens; and it legitimized claims to scientific expertise. Evidence that it fulfilled all these functions is found in the correspondence contained in this volume.

Although the scientific community was a largely inaccessible and exclusive social organization, the northern collectors who cared enough to write Smithsonian scientists usually sought recognition as their colleagues and associates. For example, Bernard Rogan Ross (1827-74), one of the most prolific correspondents and collectors, eventually belonged to the Natural History Society of Montreal, the Academy of Natural Sciences of Philadelphia, the New York Historical Society, the London Royal Geographical Society, and the Anthropological Society of London.[32] Another prodigious collector, Roderick Ross MacFarlane (1833-1920), also joined or was invited to become a member of several societies. He was a member of the American Ornithologists Union, and a Fellow of the Royal

Geographical Society, the Royal Colonial Institute, the Imperial Institute, and the National Geographical Society (U.S.).[33] The hopes and aspirations of these men were articulated many times over in the letters included in this volume; regular correspondence with Smithsonian scientists reminded metropolitan savants of their debt to the trader-collectors, at the same time that it conveyed hints as to how that debt could be repaid.[34]

THE CORRESPONDENTS

One of the most prolific and interesting of the Mackenzie River correspondents was the young naturalist sent north on behalf of the Smithsonian.[35] Robert Kennicott was obviously not a Rupert's Lander, but as an outsider he was sensitive to the ordinary as well as to the extraordinary in northern life. The narrative as well as the descriptive details of his letters provide a basis for understanding the events of the period. Moreover, Kennicott's enthusiasm for northern or "frontier" life was genuine, if selective, and he transmitted that sentiment in his correspondence. Finally, Kennicott's correspondence contains much about northern science and northern life that was missing from extant accounts. Although he had studied under some noteworthy scientists: Dr. J.P. Kirtland, an ornithologist and a founder of Western Reserve Medical College; Dr. P.R. Hoy, an amateur ornithologist from Wisconsin; Dr. D. Brainard of Rush Medical College, an expert in herpetology and physiology; and his own father, Dr. John Kennicott, a renowned horticulturalist, their instruction on the north was limited unavoidably by the paucity of published information on the region. Kennicott was familiar with George Simpson's *Overland Journey Round the World*, and Captain Cook's *Voyage to the Pacific Ocean*, but these travelogues, together with the zoological and botanical monographs and catalogues produced previously, described only a fraction of the richness of northern life.[36] As Kennicott travelled through Rupert's Land, he began collecting the missing data.

Kennicott first went to Rupert's Land in the summer of 1857. He agreed to collect zoological specimens for Northwestern University in Evanston, Illinois, and after deciding to "secure specimens from as wide a geographical range as possible," Kennicott joined the convoy of Red River carts travelling from St. Paul to the Red River Settlement.[37] He spent four months in Red River. While there he met Donald Gunn (1797-1878). Kennicott stayed with Gunn for a few days in September 1857, and at that time took the opportunity to give him some "... useful lessons for collecting and preserving all sorts of Creatures."[38] Gunn sent specimens to the Smithsonian that fall, and continued to send specimens south for the next decade.

Gunn's relationship with the Smithsonian predated Kennicott's visit, however. The old Scottish settler was one of the first residents of Rupert's Land to correspond with the Smithsonian. Extant correspondence between Gunn and the Smithsonian is dated as early as 7 June 1855, but Gunn had obviously established contact sometime earlier since this letter accompanied a meteorological register that he had kept for the Smithsonian.[39] Specimens submitted by Gunn had reached Washington by the following October, and again in April 1856.[40] By early 1856 Gunn had already asked some of his "friends to the north" to collect for the Smithsonian.

Gunn's interest in science was sincere and fairly useful to the American institution, but his dedication to natural history never equalled the intensity of his anti-HBC campaign.[42] Gunn had left Kirkwall in the Orkney Islands in 1813, joining the HBC as a labourer, but in 1822 he was discharged from his posting at Severn.[43] Gunn's services were terminated only nine years after his arrival in Rupert's Land as a result of the rationalization of the fur trade which accompanied the amalgamation of the companies in 1821.[44] He became surplus labour to the Company. He moved to St. Andrew's Parish where he farmed for ten years before taking on the positions of teacher at the Church Missionary Society school and librarian of the Red River community library.[45] He had been a settler at Red River for more than thirty-five years when he first met Robert Kennicott.

Although Gunn was one of Kennicott's first northern acquaintances, Chief Factor George Barnston (1800-83) was the young naturalist's first real convert to the Smithsonian cause. Whereas Gunn was already participating in the meteorological and natural history programs of the Smithsonian when Kennicott met him in 1857, Barnston had never even heard of the Smithsonian Institution before meeting Kennicott in 1859.[46] Barnston, like Gunn, however, was an educated man with an aptitude for science. Barnston was a native of Edinburgh, and had apparently received training as a surveyor and army engineer before leaving Scotland for North America.[47] Again, like Gunn, he was drawn to Rupert's Land by the fur trade. Barnston, however, had a long and outstanding career with the HBC. He was employed initially with the North West Company, but became a clerk with the HBC in 1821 when the companies amalgamated. In 1847, twenty-seven years after first entering the trade as an apprentice clerk with the North West Company, Barnston was promoted to Chief Factor, and he retired from the service just four years after meeting Kennicott.

Barnston and Kennicott became acquainted while travelling together from Collingwood to Fort William, but it was Barnston's keen interest in scientific activities and his curiosity about the Smithsonian expedition, rather than his credentials as a trader, that impressed the young American naturalist.[48] Barnston had already donated an entomological collection to the British Museum, and had deposited natural history specimens with McGill University, the Royal Industrial Museum of Scotland, and the Canadian Geological Museum. By 1857 he had also become a prolific contributor to scientific journals and an active member of the Natural History Society of Montreal.[49]

As Kennicott travelled north from Fort William to the Mackenzie River District, he convinced others besides Barnston and Gunn to assist in making his collections. By mid-June Kennicott had left Norway House, the HBC's major distribution depot, in the company of Julian S. Onion (1839-1907) and the Reverend W.W. Kirkby (1828-1907). Onion was a military man who had sold his lieutenant's commission with the Royal Canadian Rifles for a posting as Fort Simpson's new clerk, while Kirkby was an Anglican clergyman with the Church Missionary Society.[50] Although these men were not notable for their contributions to the Smithsonian Institution, they did submit a few specimens and wrote the occasional letter to Smithsonian scientists. A more important encounter took place at Methye Portage. The boats carrying Kennicott, the Kirkby family and Onion, reached Methye Portage on 25 July.[51] The Mackenzie River brigade arrived at the Portage on the same day. Chief Trader Bernard Rogan Ross accompanied the boats from Fort Simpson in order to oversee the exchange of the Mackenzie River fur returns for incoming trade goods and provisions.

By 1859, Ross had already spent twelve years in the Mackenzie River District.[552] He was regularly stationed at Fort Simpson, but he also spent time at Forts Norman, Liard, and Resolution.[53] Ross had, however, come by his commission with the HBC in an unconventional manner. Unlike most company men he was not recruited in Britain. He emigrated to Canada in 1843 from Londonderry, Ireland, spending his first winter in Montreal unemployed and living with a clergyman. He wanted to return home but since his chances for employment were equally poor in Ireland, his uncle, Francis Rogan, wrote an acquaintance of Governor George Simpson, asking him to recommend Ross's services to the Hudson's Bay Company.[54] That same year Ross was appointed apprentice clerk at Norway House.[55]

In total, Ross would devote twenty-four years to the HBC. His official duties never seemed adequate to fill his days, however, and throughout his career with the Company he spent a great deal of time trying to amuse himself. He was well read, wrote poetry, and took up singing and journalism. He also corresponded with scientists and scientific societies. He had, in fact, established contact with the Smithsonian at least eighteen months before Kennicott's arrival, although he only became absorbed by collecting after meeting Kennicott in person in 1859.[56] Ross submitted more than 2,200 specimens, or almost twenty percent of the total collection sent out of the Mackenzie River District during the 1860s.[57] Indeed, his contribution was significant enough to warrant special recognition. Spencer Fullerton Baird (1823-87), one of North America's most prominent ornithologists at mid-century and the Assistant Secretary of the Smithsonian Institution, named Ross's Snow Goose (*Chen rossii*) in his honour.[58]

Ross's contribution to the Mackenzie River Collection was exceeded by only one person: Roderick Ross MacFarlane. MacFarlane came from the Outer Hebrides, signing on with the company in Stornoway in 1852. He had family connections in the company, most notably his uncle Chief Factor Donald Ross, and his first posting was as an apprentice clerk at Pembina.[59] In the following outfit, he was transferred to the Mackenzie River District and made clerk in charge of Fort Rae.[60] This posting was the beginning of an extended stay in the north.

MacFarlane subsequently spent twenty years in the Mackenzie River and sixteen years in the Athabasca District. During this time, he explored the Beghula or Anderson River, and he established Fort Anderson in 1861 to facilitate trade between the HBC and the northernmost Inuit.[61]

MacFarlane knew better than anyone the potential of the Anderson, and it was his new fort that served as his operational centre when collecting for the Smithsonian. He, in fact, labelled the yearly submissions which earned him the reputation as Rupert's Land's most prodigious collector, the "Anderson Collections." MacFarlane submitted more than 5,700 specimens, or almost half of the northern specimens received by the Smithsonian during the 1860s. Like Ross, MacFarlane's contributions were recognized formally. *Falco gyrfalco* var. *sacer* was referred to as MacFarlane's gerfalcon, while an owl, *Otus asio macfarlanei* has since been named in his honour.[62] Moreover, his contributions, like those made by Ross and the other northern collectors, were referred to repeatedly in the Smithsonian Institution *Annual Report*, in ornithological monographs such as the *History of North American Birds* and *The Water Birds of North America*, and in bibliographical guides on the natural history of the Hudson's Bay and Athabasca-Mackenzie regions.[63]

Another large collection was submitted by James Lockhart (b. 1827), the person in charge of Fort Yukon from 1860 to 1863. Lockhart had been a resident of Lachine, the HBC's overseas headquarters, when he joined the service in 1849. He was only twenty-two years old when he was posted to the Lower Red River District as an apprentice clerk.[64] In 1854 he was promoted to the position of clerk, and during that same year he was also dispatched to one of the Exploratory and Search Expeditions to the Arctic.[65] The following year he was transferred to the Mackenzie River District. He was placed in charge of Fort Yukon in 1860, and promoted to Chief Trader the year after that.[66] He remained at Yukon until 1863 when he was put in charge of Fort Resolution, and he stayed there until he received furlough for the 1866-67 outfit.[67] He was relocated to Abitibi following his year's furlough, and that posting suited the pacific Lockhart.[68] Even before leaving the north, he had been anxious to move to a less taxing location. He disliked confrontation, and by 1865 the "Yankees" had already made inroads on the HBC trade along the Rocky Mountains.[69]

Lockhart sent more than 1,100 specimens, or just over nine percent of the Mackenzie River collections received by the Smithsonian; his assistant clerk, Strachan Jones, sent approximately half that number south. Jones, the son of Thomas Mercer Jones, one-time chief officer of the Canada Company in the colonies, and Elizabeth Mary Strachan, daughter of Anglican Bishop John Strachan, was a graduate of Toronto College and, according to Kennicott, a "gentleman by birth & education and a *brick* — tho' what is called a dry stick."[70] More specimens might have been expected from someone of Jones's education and social background, but despite the unimpressive size of his individual contribution, he along with Lockhart, coordinated collecting activities in the Yukon. Together they sent almost fifteen percent of all specimens out of the north.

Other HBC employees stationed north of Methye Portage also submitted specimens to the Smithsonian, but these twenty-two individuals sent only six percent of the total number sent south.[71] Most of the apprentice clerks and postmasters who collected for the Smithsonian—Thomas Swanston, William Brass, Andrew Flett, John Reid (c. 1826-95), Alexander Mackenzie, James Dunlop, Nicol Taylor (b.c. 1817) and James Flett (c.1825-99)—sent few specimens regardless of their posting; their participation was sporadic and unsustained. The most noteworthy member of this group was Charles P. Gaudet (1827-1917), one of Kennicott's favourite northern companions. In 1861, the French-Canadian trader and post supervisor—or postmaster—and Kennicott lived together for several months at Gaudet's house on the Peel River. Gaudet had been at Peel's River House since 1856, having spent the first six years of his fur trade career at Forts Resolution and Yukon.[72] He stayed at Peel's River until 1863, when he was promoted to clerk and reassigned to Fort Good Hope in the central Arctic. He was reputedly fluent in "Eskimo," and such a skill would have been useful in recruiting the Inuit visiting Good Hope to collect on behalf of the Smithsonian.

Collections made south of James Bay were also received at the Smithsonian during the 1860s. These collections never equalled those made in the Mackenzie River District, but the area was supposedly ill-suited to oological collecting. Constantin Drexler, a Smithsonian taxidermist who went to Moose Factory in the spring of 1860, wrote:

> this is the wors place for egging i have seen yet, everything is said to breed further north, and if Mr. McKenzie dooes not send me further i would had better staid at home and colected at the Smithsonian ... i will not stay at this infernal post if otherwise can be helpt, as it is shure wher ther ar no birds, ther can be no Eggs, ...[73]

Over the summer months, with the assistance of Chief Factor John Mackenzie, Drexler nevertheless managed to fill eight boxes containing between six hundred and one thousand items, including birds, fish and mammal skins, embryonic specimens, fossils, plants, shells, insects, and eggs.[74]

Specimens collected between James Bay and Michipicoten were sent south in six consecutive years. Drexler was certainly important in fostering such activity, but so were Kennicott and George Barnston. In 1859, Barnston promised Kennicott that he would ask the men in his territory to collect specimens for the Smithsonian, and several men including Chief Factor John Mackenzie, Chief Trader Joseph Gladman, R. Hamilton, B. Smith and Colin Rankin sent specimens either as a result of, or through, George Barnston.[75]

Barnston's importance to Smithsonian field work clearly differed from most northerners who corresponded with metropolitan scientists. His empirical contributions may have been dwarfed by those of MacFarlane, Ross, Kennicott, and Lockhart, but he was instrumental in establishing a Smithsonian presence south of James Bay. He began recruiting collectors for the Smithsonian a full year before Drexler went into Rupert's Land. A less dramatic expression of the contributions made by Barnston also surfaces in his willingness to assist Kennicott's travels through Rupert's Land. While at Sault Ste. Marie, Barnston arranged a "certificate of deposit" for £142.12.5 sterling which allowed Kennicott to "raise money anywhere in the Fur Co.'s possessions." He also gave the young American free passage on the HBC's canoes from Fort William to Norway House.[76]

Barnston was a facilitator as well as a collector. So too was William Mactavish (1815-70). Although Mactavish had been a company man since 1833 when his uncle got him an apprenticeship with the HBC, he became an important political figure in the Red River Settlement when he was made the Governor of Assiniboia in 1858.[77] When Governor Simpson died in 1860, Mactavish acted as interim governor of the HBC until 1862 when Alexander Grant Dallas (1816-82) was appointed to the overseas governorship. Mactavish became governor in his own right in 1864, however, when Governor Dallas resigned. He was then responsible for both civil and corporate affairs in Rupert's Land, and was a very powerful man during the period in which the Smithsonian conducted fieldwork in the north.

Mactavish's interest in science was especially appreciated after Dallas took over the HBC governorship. According to the authors of the letters contained in this volume, scientific activities were thwarted in the period between 1862 and 1864 because Governor Dallas dismissed the importance of collecting. He tended to view it as little more than a form of entertainment; it was a fine way to fill idle hours or to acquire exotica from distant and foreign parts. Numerous complaints about Dallas's insensitivity and outright antipathy to scientific collecting surface in the correspondence included in this volume. His cavalier attitude to the collections and collecting activities of his northern employees demonstrated his indifference to the role that lay collectors could play in serious science, and it offended some collectors. Mactavish, on the other hand, was himself interested in science, and actually sent specimens to the Smithsonian on four occasions.[78] As governor, he was also able to expedite communications and transportation between Rupert's Land and Washington. His attention ensured that specimens were shipped out of Rupert's Land regularly and carefully; it also expedited the transmission of Smithsonian freight—the books, instructions, instruments, and preservatives—to points in the north. Such support and cooperation were instrumental to the success of the Smithsonian collecting program in Rupert's Land.

Between 1862 and 1864, the Dallas governorship reminded northern collectors and Smithsonian scientists of the importance of well-placed individuals to field programs. Robert Kennicott was, of course, one such person. Although Kennicott was an able and energetic field naturalist, he needed help to amass the kind of collections expected at the Smithsonian. He was, however, quite capable of motivating others to work on behalf of "science." Indeed,

Kennicott's success at convincing local residents to do fieldwork, and his ability to then convert them into highly skilled collectors and taxidermists, were two of his most valuable talents. Kennicott's devotion to natural history was itself inspiring, and many of his recruiting successes rested on example. Eighteen hour days in the field, and an equally tiresome pace associated with keeping field notes, measuring specimens, and attending to the preservation and packing of collections, convinced northerners of the importance of his expedition and motivated many to take up science.

Northerners were also convinced of the importance of collecting for the Smithsonian by Spencer Fullerton Baird, the Assistant Secretary of the American institution and one of the

Spencer Fullerton Baird (1823-87), Assistant Secretary of the Smithsonian Institution, circa 1860. Smithsonian Institution Photo #64750.

most respected naturalists of his generation.[79] In 1850, after spending nearly five years as a professor of natural history at Dickinson College in Carlisle, Pennsylvania, he became the first Assistant Secretary of the Smithsonian Institution in charge of Natural History. Between 1850 and 1878, Baird supervised the acquisition of the collections needed for taxonomic and systematic research; he compiled and wrote several zoological catalogues, monographs, and scholarly articles; and he coordinated the production of a series of instructional pamphlets that outlined proper procedures for field workers. Administrative and editorial duties were also part of Baird's work as Assistant Secretary. He managed the exchange of books and journals between North American and European scientific institutions; he edited numerous reports on scientific data obtained by government exploring expeditions; he edited a variety of scientific journals, including the Smithsonian Institution *Annual Report* and the *Annual Record of Science and Industry*; and through his position at the Smithsonian he facilitated the printing and dissemination of the transactions of the Washington Philosophical Society, the Anthropological Society, and the Biological Society. In recognition of his contributions to North American science, Baird was made the first Commissioner of Fisheries in 1871, and he succeeded Joseph Henry (1797-1878) as the second Secretary of the Smithsonian in 1878.

Baird was also, as attested by the contents of this volume, an inveterate letter-writer. His output was extraordinary; he wrote, for example, 3050 letters in 1860 alone.[80] He responded to each and every inquiry, and his correspondence with collectors contained praise for past efforts as well as encouragement for future endeavours. His willingness to write these individuals was also at least partially fuelled by an interest in his collectors that was genuine and insatiable. Kennicott noted Baird's "fondness for gossip," and the letters contained in this volume would certainly satisfy even the most curious of individuals.

Although Baird was the primary recipient of the correspondence out of Rupert's Land, northerners also wrote Joseph Henry, an important physicist who taught at Princeton University before becoming the first Secretary of the Smithsonian Institution in 1846. Baird's adolescent daughter, Lucy (1848-1913), received letters from Kennicott, and Kennicott himself received numerous letters from his northern acquaintances after he returned home to the United States in 1862. Kennicott died prematurely in 1866 while exploring the Russian Territories (Alaska) as the Smithsonian representative on the Western Union Telegraph Company expedition, but many northerners wrote him in the intervening years. Kennicott received letters from Ross, MacFarlane and Lockhart, both at his home—The Grove—in Illinois, and in Washington where he spent several months in 1863 and 1864 sorting and identifying the northern collections. Many of these letters are included in this volume. They contain some of the most personal and interesting passages to be found in the entire collection, and provide some unusually candid comments about northern life and the fur trade.

EDITORIAL PROCESS

Although the text of these letters is usually accessible, some annotation is unavoidable. I have, however, endeavoured to editorialize sparingly. Headnotes and an appendix have replaced extensive and repetitious footnoting, and the maps contained in the Introduction are substituted for written descriptions of obscure HBC posts and geographical points. Northern correspondents regularly, if not always correctly, used scientific nomenclature, and the common names of the Latinized binomes used by the scientific community in the mid-nineteenth century are given in the Appendix. Unless otherwise cited, explanatory notes are based on secondary sources, especially Canadian, American and British biographical dictionaries. Descriptions of lesser known persons, particularly HBC employees, are based on the primary sources listed in the Biographical Files at the Hudson's Bay Company Archives. Some monographs have also proved useful in constructing the notes, specifically: W.H. Dall, *Spencer Fullerton Baird* (1915); J.J. Hargrave, *Red River* (1871); S. Van Kirk, '*Many Tender Ties*' (1980); J.S.H. Brown, *Strangers in Blood* (1980); A. Gunther, *A Century of Zoology* (1975); and W. Goetzmann, *New Lands, New Men* (1986). As well, I consulted *The Beaver* while researching people and places referred to by northern correspondents.

For the most part, the letters reprinted here have been reproduced unchanged. Textual editorializing has been kept to an absolute minimum so as to adhere, as closely as possible, to

Joseph Henry (1797-1878), Secretary of the Smithsonian Institution, 1846-78. Smithsonian Institution Photo #82-3249.

a literal transcription of the letters. Sometimes the grammar and syntax is unusual, or simply incorrect, but these stylistic idiosyncrasies have been preserved. HBC clerks were often guilty of overusing the upper case, but their capitalization practices have been left intact. Erroneous or unconventional spellings have been left as written originally, but words or passages stroked out by the author have not been replicated. I have not inserted *[sic]* after unusual or erroneous spellings because unconventionality was more often the case than not in the orthography of everyday and scientific words uses by northerners. Editorial comments do appear within square brackets, however, and they are italicized. Italics also indicate emphasis found in the originals. The dashes which commonly accompanied periods or commas in nineteenth century prose have, however, been removed. Dashes used instead of periods or commas have also been replaced by the appropriate punctuation.

Although many of the letters are riddled with run-on sentences, I have usually left these as written. Some of the correspondence was almost incomprehensible without intervention, however. The letters written by William Mactavish were essentially one long sentence, having neither punctuation nor capitalization, and both were inserted silently. Robert Kennicott's letters were often written as if they were hasty notes or memos—often a series of points jotted down as they occurred to him—and where his letters could be made more lucid by simply adding the occasional comma or period, punctuation has been added silently. Donald Gunn's letters posed a similar problem, and I have treated them in the same manner as those written by Kennicott and Mactavish. Punctuation and capitalization has only been inserted to make the letters more readable; no attempt has been made to correct faulty grammar and syntax, or to expand or modernize the text.

An attempt has been made to reproduce these letters very much as they were written over one hundred and twenty-five years ago, and so they have been arranged to reflect, as accurately as possible, the chronology and circumstances of their creation. Mail intended for Canadian, European or United States destinations only left Rupert's Land twice a year. Although mail could be sent out of Red River more frequently, for the most part postal services in the north were biannual. This pattern was certainly true for the region from which most of the correspondence reprinted here was generated. Aside from letters originating south of the Mackenzie River District, correspondence out of Rupert's Land left in the fall or spring "packet."[81]

Because the spring or summer packet left Fort Simpson in mid-June, letters from Fort Yukon, for example, had to be written before the boat left that post on 1 June. The boat going south from Fort Yukon made it to Fort Simpson just before the winter packet inward, which had left Fort Garry the previous December, arrived. Letters arriving after mid-June could only go out six months hence, when the fall or winter packet left for the south in December. Outgoing mail from even the most distant post in the district therefore had to be in Fort Simpson by the end of November. The constraints of the HBC packet system emerge in a pattern that is replicated herein. In order to depict the sequence of developments within the Rupert's Land - Smithsonian collecting network, I have arranged the correspondence chronologically. Correspondence generated outside of the northern packet system has been inserted where most appropriate.

ACKNOWLEDGEMENTS

I am indebted to many individuals and organizations, but I most especially thank the Smithsonian Institution Archives for allowing me to publish letters and photographs from their collections. The Manitoba Record Society also played an important role in this project. Jack Bumsted, the General Editor of the Society, deftly balanced enthusiasm and encouragement with the mechanics of editing so that I always felt that my project was as important as the product. Society Directors, Angela Davis and Barry Ferguson, graciously agreed to read the manuscript, and I thank them for their comments. The Record Society generously made funds available to cover the costs of typing the manuscript and for reproducing photos from the Provincial Archives of Manitoba, and I am most grateful. Archivists in Manitoba and Washington have offered their expertise and services unreservedly; I thank Bill Deiss and Susan Glenn at the Smithsonian Institution Archives, and Judith Beattie at the Hudson's Bay

Company Archives, Provincial Archives of Manitoba. I would like to thank Victor Lytwyn of the Geography Department at the University of Manitoba for drawing the maps included in this volume. His skills as a cartographer and his expertise as a historical geographer were put to the test with some of the geographical references found in the HBC-Smithsonian correspondence. Finally, and most gratefully, I would like to thank Karen Morrow for being both palaeographer and proof-reader as she typed photocopied originals and my equally indecipherable annotations. Errors are, of course, my responsibility.

NOTES

1 See M.L. Tyrwhitt-Drake, "David Douglas," *Dictionary of Canadian Biography [DCB]*, vol. 6, pp. 218-220; Grace Lee Nute, "A Botanist at Fort Colvile," *The Beaver*, Sept. 1946, pp. 28-31; S. Zeller, *Inventing Canada: Early Victorian Science and the Idea of a Transcontinental Nation* (Toronto: University of Toronto Press, 1987), p. 212; and G. Thomas, "The Smithsonian and the Hudson's Bay Company," *Prairie Forum*, x, 2 Fall 1985, p. 285.

2 Ethnographic observations were becoming scientific "facts" in the mid-nineteenth century, and such accounts were essential to the emerging social sciences. See Elman Service, *A Century of Controversy: Ethnological Issues from 1860 to 1960* (Orlando; Academic Press, Inc., 1985); Mary Louise Pratt, "Fieldwork in Common Places," and Vincent Crapanzano, "Hermes Dilemma: The Masking of Subversion in Ethnographic Description," in *Writing Culture: The Poetics and Politics of Ethnography,* eds. James Clifford and George E. Marcus, Experiments in Contemporary Anthropology, A School of American Research Advanced Seminar (Berkeley: Universiyt of California Press, 1986).

3 One third of the charter members of the HBC were Fellows of the Royal Society, see E.E. Rich, *The History of the Hudson's Bay Company,* vol. 1, as well as vol. 5 of the Hudson's Bay Record Soceity publications, *Minutes of the Hudson's Bay Company, 1671-1674,* pp. xxvi-xxviii; and R.P. Stearns, "The Royal Society and the Company," *The Beaver*, June 1945, pp. 8-13.

4 Stearns, "The Royal Society and the Company," pp. 9-12.

5 Knight was briefly made Governor-in-Chief of Rupert's Land, 14 July - 13 September, 1717. See Glyndwr Williams, ed., *Andrew Graham's Observations on Hudson's Bay, 1767-91*, with an Introduction by Richard Glover, vol. 27 (London: The Hudson's Bay Record Society, 1969), endnote 1, p. 243. Also see, Ernest S. Dodge, "James Knight," *DCB*, vol. 2, pp. 318-20.

6 Rich, *The History of the Hudson's Bay Company, 1670-1870*, vol. 1, pp. 446-447.

7 E.E. Rich, ed., *James Isham's Observations on Hudsons Bay, 1743, and Notes and Observations on A Book Entitled A Voyage to Hudsons Bay in the Dodds Galley, 1749,* Assisted by A.M. Johnson, with an Introduction by Rich, vol. 12, Hudson's Bay Record Society (Toronto: The Champlain Society, 1949), p. xlvii-lii.

8 Arthur Dobbs, *An Account of the Countries adjoining to Hudson's Bay, in the North-West Part of America: Containing a Description of their Lakes and Rivers, the Nature of the Soil and Climates, and their Methods of Commerce, &c., Shewing the Benefit to be made by settling Colonies, and opening a Trade in these Parts whereby the French will be deprived in a great Measure of their Traffick in Furs, and the Communication between Canada and Mississippi be cut off* (London: J. Robinson, 1744), p. 2.

9 Rich, *James Isham's Observations on Hudson's Bay*, p. xc-xcix.

10 John Richardson, William Swainson and Reverend William Kirby, *Fauna Boreali-Americana, or the Zoology of the Northern Parts of British America, Contaiing Descriptions of the Objects of Natural History Collected on the Late Northern Land Expeditions, Under Command of Captain Sir John Franklin,* Three vols., Part I - *The*

Quadrupeds, Part II - *Birds,* Part III - *Fishes* (London: John Murray, 1829-36). See Part II, pp. ix-x. Also see James L. Baillie Jr., "Naturalists on Hudson Bay," *The Beaver,* Dec. 1946, pp. 36-39.

11 Ruth Dawson, "Collecting with Cook: The Forsters and their Artifact Sales," *Hawaiian Journal of History* 13 (1979), 5-16. See Edward A. Preble's bibliography for reference to Forster's usage of HBC specimens. *North American Fauna,* No. 22, *A Biological Investigation of the Hudson Bay Region,* Department of Agriculture, Division of Biological Survey (Washington: Government Printing Office, 1902), p. 28.

12 Richard Glover, Introduction to *Andrew Graham's Observations on Hudson Bay, 1767-91,* ed. Glyndwr Williams, vol, 27 (London: The Hudson's Bay Record Society, 1969), pp. xiv-xv and pp. xxii-xxiii.

13 Stearns, "The Royal Society and the Company," p. 12.

14 See Stearns for Company contributions to the Royal Society in this early period, pp. 8-13.

15 G. Williams, *Andrew Graham's Observations on Hudson's Bay, 1767-91,* p. 357.

16 Hearne, *A Journey from Prince of Wales Fort, in Hudson's Bay, To the Northern Ocean, Undertaken by Order to the Hudson's Bay Company, For the Discovery of Copper Mines, A NorthWest Passage &c., In the Years 1769, 1770, 1771 & 1772.* (London: A. Strahan and T. Cadell, 1795; reprint ed., Toronto: Macmillan Co. of Canada Ltd., 1958), pp. xxxvi-ii, footnote 12 especially; xli and xlii.

17 A letter written by the Home Board to Governor Thomas Thomas in Rupert's Land is the earliest found record of their instructions regarding the submission of Reports and Journals by HBC employees overseas. Hudson's Bay Company Archives [H.B.C.A.], Provincial Archives of Manitoba [PAM], A.6/18, pp. 149-213.

18 A series of articles on Arctic exploration appeared in *The Beaver* in 1969. See: John E. Caswell, "The Sponsors of Canadian Arctic Exploration," Spring 1969, pp. 4-13; Summer 1969, pp. 38-45; Autumn 1969, pp. 25-33; and Winter 1969, pp. 44-53. More recently two popular accounts have traced Arctic exploration, see Daniel Francis, *Discovery of the North: The Exploration of Canada's Arctic* (Edmonton: Hurtig Pubs., 1986) and Pierre Berton, *The Arctic Grail: The Quest for the North West Passage and the North Pole, 1818-1909* (Toronto: McClelland & Stewart, 1988). A more scholarly approach to the topic can be found in Trevor Levere, "Science and the Canadian Arctic, 1818-76, from Sir John Ross to Sir George Strong Nares," *Arctic,* vol. 41, no.2 (June 1988), 127-37.

19 Alan Cooke and Clive Holland, *The Exploration of Northern Canada, 500-1920: A Chronology* (Toronto: The Arctic Press, 1978), pp. 151-53. Also see John Edwards Caswell, "United States Scientific Expeditions to the Arctic 1850-19009," Ph.D. 1951, Stanford University.

20 Resolution 87 of Minutes of Northern Council, 1851. Resolution 88 directed HBC officers to keep an account of the costs expended with regard to the British search expeditions, and to forward this account to Lachine for settlement with the British government. H.B.C.A., PAM, B.239/k/3, p. 21

21 See Thomas Simpson, *Narrative of the Discoveries on the North Coast of America Effected by the Officers of the Hudson's Bay Company During the Years 1836-39* (London: Richard Bentley, 1843), p. 6 Also see J.M. Wordie and R.J. Cyriax, Introduction to *John Rae's Correspondence with the Hudson's Bay Company on Arctic Exploration, 1844-1855,* ed. E.E. Rich, Assisted by A.M. Johnson, vol. 16 (London: The Hudson's Bay Record Society, 1953).

22 John Rae, *Narrative of an Expedition to the Shores of the Arctic Sea, in 1846 and 1847, with Maps* (London: T. & W. Boone, 1850), p. 15.

23 John Ross, *A Voyage of Discovery Made Under the Order of the Admiralty in His Majesty's Ships 'Isabella' and 'Alexander', For the Purpose of Exploring Baffin's Bay, and Inquiring into the Probability of a North-West Passage* (London: John Murray, 1819), pp. v-xcliv.

[24] John Franklin, *Narrative of a Second Expedition to the Shores of the Polar Sea, In the Years 1825, 1826, and 1827, Including an Account of the Progress of a Detachment to the Eastward by John Richardson* (Lonton: John Murray, 1828), p. vii.

[25] The botanical specimens were described and classified by William J. Hooker and the first volume of his work appeared in 1833.

[26] George F. G. Stanley, ed., *John Henry Lefroy: In Search of the Magnetic North, A Soldier-Surveyor's Letters from the North-West, 1843-1844* (Toronto: The Macmillan Co. of Canada Ltd., 1955). Also see Zeller, *Inventing Canada*, pp. 125-44.

[27] See Morris Zaslow, *Reading the Rocks: The Story of the Geological Survey of Canada, 1842-1972* (Ottawa: The Macmillan Co. of Can. Ltd., in assoc. with the Department of Energy, Mines and Resources, and Information Canada, 1975), pp. 82-127. Also see Zeller, *Inventing Canada,* p. 14 and 97-98.

[28] Debra Lindsay, "Science in the Sub-Arctic: Traders, Trappers and the Smithsonian Institution, 1859-70," Ph.D. 1989, University of Mantioba.

[29] Cooke, and Holland, *The Exploration of Northern Canada*, pp. 220-308.

[30] See Zaslow on the role of the Canadian Geological Survey in northern exploration, especially Chapters 5-11 in *Reading the Rocks.*

[31] Information on the function of scientific societies can be found in T.W. Heyck, *The Transformation of Intellectual Life in Victorian England* (New York: St. Martin's Press, 1982), pp. 57-62; Richard A. Jarrell, "The Social Functions of the Scientific Society in Nineteenth Century Canada, " in *Critical Issues in the History of Canadian Science, Technology and Medicine,* Jarrell and Arnold E. Roos, eds. (Thornhill & Ottawa: HSTC Publications, 1983), pp. 31-44, and Sally Gregory Kohlstadt, *The Formation of the American Scientific Community: The American Association for the Advancement of Science 1848-60* (Champaign, Illinois: University of Illinois Press, 1976).

[32] Debra Lindsay, "The Hudson's Bay Company-Smithsonian Connection and Fur Trade Intellectual Life: Bernard Rogan Ross, a Case Study," in *Le Castor Fait Tout,* Selected Papers of the Fifth North American Fur Trade Conference, 1985, eds., Bruce Trigger, Toby Morantz and Louise Dechene (Montreal: Lake St. Louis Historical Society, 1987), p. 609.

[33] R. MacFarlane, "A Brief Sketch of the Life and Services of Retired Chief Factor R. MacFarlane, 1852-1913," Extracted from the third volume of *The Story of Manitoba,* ed. Schofield. PAM MG14 C23 #47.

[34] See W.A. Deiss on Assistant Secretary S.F. Baird's reaction to incoming correspondence. "Spencer F. Baird and his collectors," *Journal of the Soceity for the Bibliography of Natural History* (1980) 9(4), pp. 635-45.

[35] Biographical information on Robert Kennicott, and his father John, has been obtained from James A. James, ed., *The First Scientific Exploration of Russian America and the Purchase of Alaska* (Chicago: Northwestern University, 1942), pp. 1-6. and from Keir Sterling's Introduction to *American Natural History Studies: The Bairdian Period,* The Natural Sciences in America (New York: Arno Press, 1974), no pagination. Information on Robert Kennicott can also be found in Donald Zochert, "Notes on a Young Naturalist," *Audubon,* March, 1980, pp. 34-47; Grace Lee Nute, "Kennicott in the North," *The Beaver,* Sept. 1943, pp. 28-32 and in the *Biographical Dictionary of American Science, The Seventeenth Through the Nineteenth Centuries [BDAS],* Clark A. Elliott, ed. (London; Westport, Conn.: Greenwood Press, 1979), p. 145. Also see "Jared Potter Kirtland," *BDAS*, pp. 147-48.

[36] Information about Kennicott's education and curricula is found in J.A. James, *The First Scientific Exploration of Russian America,* p. 1, f.n. Cook's book is incorrectly entitled *Vancouver,* however.

[37] Ibid, p. 4.

[38] Donald Gunn to Baird, 26 Nov. 1857, Smithsonian Institution Archives [SIA], Record Unit

[RU] 305, Reel 6.

[39] Gunn to Baird, SIA, RU 52, Box 7, vol. 2, p. 329.

[40] Baird wrote Gunn acknowledging receipt of his specimens on 11 Oct. 1855 and 10 April 1856, SIA, RU 57, Box 2.

[41] Gunn to Baird, 20 Feb. 1856, SIA, RU 305, Reel 6.

[42] See, for example, Gunn to Baird, 11 Jan. 1858, SIA, RU 305, Reel 7.

[43] H.B.C.A., PAM, A.16/37, fo.110.

[44] Servant's Accounts, H.B.C.A., PAM, B.239/g/2.

[45] L.G. Thomas, "Donald Gunn," *DCB*, vol 10, p. 324.

[46] Kennicott to Biard, 17 Nov. 1859, SIA, RU 7215, Box 13.

[47] Jennifer S.H. Brown and Sylvia Van Kirk, "George Barnston," *DCB*, vol, 11, pp. 52-53.

[48] Kennicott to Baird, 7 May 1859, SIA, RU 7002, Box 27.

[49] Brown and Van Kirk, "George Barnston," *DCB* vol. 11, pp. 52-53; George A. Dunlop and C.P. Wilson, "George Barnston," *The Beaver,* Dec. 1941, 16-17; and Bernard Rogan Ross Notebook, SIA, RU 7221.

[50] Biographical information on Onion can be found in Charles Camsell, *Son of the North* (Toronto: The Ryerson Press, 1954), pp. 2-4; Minutes of Council, Northern Department, Resolution 74, 1862, H.B.C.A., PAM, B.239/k/13; and the Alexander Morris Papers, PAM, MG12 B1 #1934. For information on Kirkby see Thomas C.B. Boon, "William West Kirkby, First Anglican Missionary to the Loucheux," *The Beaver*, Spring 1965, pp. 36-43.

[51] Kennicott to Baird, 27 July 1859, SIA, RU 7215, Box 13.

[52] See Hartwell Bowsfield, "Bernard Rogan Ross," *DCB*, vol. 10, p. 629. Also see Lindsay, "The Hudson's Bay Company-Smithsonian Connection," pp. 587-617.

[53] H.B.C.A., PAM, B.239/k/3.

[54] Francis Rogan to Isaac Calhorn, 27 March 1843, H.B.C.A., PAM, E.31/1.

[55] H.B.C.A., PAM, B.239/k/2.

[56] The first extant correspondence between Ross and the Smithsonian is dated 28 Nov. 1859 SIA, Hudson's Bay Company Correspondence Collection [HBC Corr Coll], Folder 36.

[57] See "List of Species Collected at Fort Simpson, 1860 and 1861" by B.R. Ross, H.B.C.A., PAM, E.37/13, and "Collected Notes, Lists and Catalogs on Birds," SIA, RU 7215, Box 13, Robert Kennicott; Box 29, Folder: B.R. Ross. Also see the accession records of the Anthropology Department at the Smithsonian Institution, SIA, RU 6990T and Computer Printout, Indent. MNH4 122G113, MNH-ANN.

[58] S.F. Baird, T.M. Brewer and R. Ridgway, *The Water Birds of North America,* Natural Sciences in America, ed., Keir B. Sterling (Boston: Little, Brown & Co., 1884; reprint ed. New York: Arno Press, 1974), p. 445.

[59] H.B.C.A., PAM, A.10/82b, fos. 187-197.

[60] H.B.C.A., PAM, B.239/k/3, p. 48.

[61] "Retired Chief Factor R. MacFarlane," PAM,. MG14 C23 Box 2 #47, p. 5.

[62] The common name is referred to in Baird, Brewer and Ridgway, *History of North American Birds: Land Birds,* 3 vols., First pub. 1874 (Boston: Little, Brown & Co., 1905), p. 115, and the scientific name can be found in any field guide, for example, R.T. Peterson, *A Field Guide to Western Birds,* second ed. (Boston: Houghton Mifflin Co., 1961).

[63] See monographs by Baird, Brewer and Ridgway, *History of North American Birds* and

The Water Birds of North America. Bibliographies include "A Biological Investigation of the Hudson Bay Region," *North American Fauna,* no. 22, and "A Biological Investigation of the Athabaska-Mackenzie Region," *North American Fauna,* no. 27, U.S. Department of Agriculture, Bureau of Biological Survey (Washington: Government Printing Office, 1908) by Edward Alexander Preble.

64 Simpson to Hargrave, 28 June 1849, H.B.C.A., PAM, B.239/c/5.

65 H.B.C.A., PAM, E.15/11.

66 H.B.C.A., PAM, B.239/k/13.

67 See correspondence from Lockhart to Baird for details of the traders whereabouts while on furlough. He spent most of the year visiting his sister's family in Montreal, and visiting his brother's family in Whitby, Canada West. SIA, HBC Corr Coll, Folder 26.

68 Lockhart to Baird, 9 April 1867, SIA, HBC Corr Coll, Folder 26.

69 Lockhart to Kennicott, 26 June 1865, SIA, HBC Corr Coll, Folder 26.

70 See Roger D. Hall, "Thomas Mercer Jones," *DCB*, vol. 9, pp. 415-17; Jones to Baird, sent from Toronto and received at the Smithsonian on 9 April 1867, SIA, HBC Corr Coll, Folder 24; and W.W. Kirby *[sic]*, "A Journey to the Youcan, Russian America," Smithsonian Institution *Annual Report[SIAR],* 1864, p. 417. The passage cited is from a letter written by Kennicott to Baird, 23 June 1861, SIA, RU 7215, Box 13.

72 See Appendix A, "Northern Contributions to the Smithsonian, 1859-71," in Lindsay, *Science in the Subarctic* (Washington, D.C.: Smithsonian Institution Press, forthcoming Autumn 1992).

72 Information on Gaudet has been derived from the H.B.C.A. Search File, "Misc. G Folder People," from an article written by one of Gaudet's descendants, J.L. Gaudet, "Chief Trader Charles Philip Gaudet," *The Beaver*, Sept. 1935, pp. 45, and from a letter written by J.L. Gaudet to R.R. MacFarlane 3 April 1918. MacFarlane forwarded Gaudet's letter to Edward Alexander Preble, who was compiling biographical data on the HBC collectors, at the Smithsonian. SIA, RU 7252, Box 3, Folder 13. There is some discrepancy between these sources and the HBC Minutes of Council regarding the year Gaudet joined the company. He is registered as a postmaster in 1851, at the lowest rate of pay of £25. H.B.C.A., PAM, B.239/k/12, p. 322. Also see Kennicott's description of life at Gaudet's post in "A Rubbaboo Journal for Friends at Home," in J.A. James, *The First Scientific Exploration of Russian America*, pp. 85-135.

73 Drexler to Baird, 26 May 1860, SIA, HBC Corr Coll, Folder 16.

74 The imprecision in the specimen counts is due to the rather vague manner in which Drexler recorded the items on the invoice. See "Invoice of contents of 8 boxes addressed *[sic]* to the Smithsonian Institution, Washington, D.C.U.S.", SIA, HBC Corr Coll, Folder 16.

75 *SIAR*, 1860-65.

76 See Kennicott to Baird, 7 May 1859 and 9 May 1859, SIA, RU 7002, Box 27.

77 N. Jaye Goossen, "William Mactavish," *DCB*, vol. 9, pp. 529-32.

78 *SIAR*, 1860, 1863, 1864, 1865.

79 Biographical information on Baird comes from a variety of sources, including: George Brown Goode, "The Three Secretaries," in *the Smithsonian Institution, 1846-1896, The History of Its First Half Century* (Washington, 1897), pp. 115-234; William Healey Dall, *Spencer Fullerton Baird, A Biography* (Philadelphia & London: J.B. Lippincott Co., 1915); William A. Deiss, "Spencer F. Baird and his Collectors," *Journal of the Soceity for the Bibliography of Natural History* (1980)9(4), pp. 635-645; and W.A. Deiss, "The Making of a Naturalist: Spencer F. Baird, The Early Years," in *From Linnaeus to Darwin, Commentaries on the History of Biology and Geology* (London: Society for the History of Natural History, 1985).

[80] Deiss, "Spender F. Baird and his collectors," pp. 638-39.

[81] See Kennicott to Baird, 27 June 1859, for an extensive discussion of nothern transportation and communication timetables and routes. SIA, RU 7215, Box 13. Also see letters from James R. Clare to Baird, 1866, for further information. SIA, HBC Corr Coll, Folder 8.

Selected Letters from Rupert's Land to the Smithsonian Institution, 1856-1868

1. DONALD GUNN to SPENCER BAIRD

Red River Settlement Feb 20th/56
[SIA, RU 305, Reel 6]
[While the following letter is the first reprinted here, it was not Gunn's first communication with the Smithsonian. Baird received a letter from Gunn dated 7 June 1855.]

Dear Sir,

After a long delay, for which, I crave your indulgence, I beg to acknowledge the receipt of your communication of Oct. last which came to hand by the Novr. mail. I have not hitherto been able to get any specimens of the mamalia so much wished for by you. Mice are very scarce this Winter Owing, perhaps, to the unusually Wet summer which we have had, or it may be from the effects of this extra-ordinary cold winter which prevents them moving freely on the surface of the snow, however I will not lose sight of your request, and I shall find the greatest pleasure in seconding the laudable objects of the Smithsonian Institution in however humble a degree I may be able to do so. All spirits in this place is a monopoly of the H. Bay Coy. consequently very dear: Alcohol fourteen shillings sterling per liquid pound, common rum twenty shillings per Gall. When fortunate enough to get any specimens of Mice I will put them in Rum. There is a variety of squirrels. I shall endeavour to procure some and preserve them by skinning. I beg to observe that lint is not in any quantity to be had here—in stuffing birds we have sometimes used dry moss. However, I think moss is not very safe, as small Insects often lay their Eggs in it which the least moisture would vivify—after which the young creatures would prey on the skins. Pray if convenient send some lint. I am desirous of collecting specimens of birds. We have a few which pass the winter with us and a numberless Variety of birds of passage, and many which pass their winters in the Warmer regions of the South take up their abode here during the summer months. We have only one species of the Hare tribe, viz the Common Rabbit, which is to be met with some years in very great numbers throughout all the regions around Hudsons Bay. In

Donald Gunn (1797-1878) Donald Gunn Collection N 9976 Provincial Archives of Manitoba.

the waters of Red River and Lake Winnipeg there is no Trout. Whitefish of the very best quality are in great abundance. Sturgeon weighing from a few pounds Wt to Eighty or more. Some of the largest Measure five feet in length—two varieties of perch, three varieties of succors, pikes, sunfish, catfish and Gold Eyes. You will confer a favour by directing me how I can forward any collection, that I may be able to make. I think the carriage ride be expensive and I do not know whether the Institution is prepared to meet such demands. The three registers which I forwarded to your institution during the summer and Harvest cost me 2/3° sterling for postage, to avoid which in future I should forward when an opportunity offers when no postage will be required. In Company with this forward five registers which brings up my observations to the beginning of the *[previous?]* Month. Some of the Red Lake missionaries are here on a visit. They intend leaving tomorrow for their home, and have kindly offered to take any communication for the Institution along with them.

I have written to some friends to the North for their assistance in collecting specimens for your work. There are many varieties of Mice even on the frozen shores of Hudson's Bay and in the Forests surrounding the innumerable streams falling into it.

I have not as yet heard from any of these Gentlemen, but I trust that they will take deep interest in the advancement of Natural History and contribute their Mite.

Would specimens of our Rocks Earth etc. be acceptable. If so please let me know.

Hoping to hear from you when convenient. I Remain, Very truly Yours,

DONALD GUNN

2. DONALD GUNN to SPENCER BAIRD

Red River Settlement June 2nd 1857

[SIA, RU 7215, Box 10]

[Charles Cavalier was appointed customs officer at Pembina when an office was established in 1851. In this position he officially controlled mail service out of Rupert's Land, via the United States, until 1855 when he relocated to St. Joseph and then to St. Boniface. When he returned to Pembina in 1863 he built the first post office in North Dakota, and although Norman Kittson was the first postmaster at Pembina, Cavalier occupied the position for twenty-one years. The Richardson reference comes from *Arctic Searching Expedition, a Journal of a Boat Voyage through Rupert's Land and the Arctic Sea, in Search of the Discovery Ships Under Command of Sir John Franklin. With an Appendix on the Physical Geography of North America,* 2 vols. (London: Longman, Brown, Green and Longmans, 1851).]

My Dear Sir,

I have great pleasure in acknowledging the receipt of your much esteemed favors of 9th Augt and 13th Oct/56, and at the same time, I beg that you will bear with me, for not answering the same at an earlier date. My reason for not writing was that, I had nothing of importance to communicate, and I did not like to trouble you, or take up your time with unimportant, if not altogether useless correspondence. The Volume containing the *Report* of the Institution for 1854, which you had the great kindness to forward to me, I received for which I tender you my most sincere and heartfelt thanks for the handsome Manner in which you have been pleased to remember me.

I am sorry that the specimens from Nelson River have not been done well in the beginning. Many of them had been prepared by the Indians and stuffed with Carriboes hair which has had a bad effect on some of the Mice, a few of them are losing the hair and I fear that they will not answer for setting up in a cabinet, but they will serve I trust to take drawings from. I have packed up the few specimens which I have been able to procure, and I am sorry that they are so few, and I fear will be of little Interest to you. In a small tin Jar I have put seven Mice and a Lamprey. I trust that the rum in which they are is sufficiently strong to keep them sound. I send 4 skinned mice, two of them rather injured. They were all procured in this Vicinity, and I have given a very imperfect description of them in another sheet. There are other kinds of Mice in this District of which I have not yet been able to procure any specimens. Altho I have applied to several Indians to try to bring me some of these, I have learned this winter that the pouched rats are found in our locality, within the bounds of this Colony, and I shall try to procure some of them.

I have written to some of my Friends, in different parts of the Territory, desiring them to procure specimens of the smaller Mammals,

Cavalier's Post Office. Mail and specimens leaving Rupert's Land for the United States were routed through Pembina, and Cavalier's Post Office was a major point on the Smithsonian-Mackenzie River collecting network. George H. Gunn Collection, 64. Provincial Archives of Manitoba.

viz Mice, squirrels, etc., etc. and the Skeletons of the Wolverine. I am not able to say how far any of them may be able or feel disposed to assist - as I have not heard from any of them Since I wrote to them. But if any of them will forward anything I cannot receive the same till July, and I expect I will be able to forward the same in the fall of the Year.

In one of your letters you express or desire to know whether the brown bear be an accidental Variety or a distinct specie. These animals in what is called the Hudson's Bay Territories are so few that I can hardly venture to think them a distinct variety. Yet they differ in many respects from the common black bear. The Indians represent them as more ferocious, their fur is much finer and their skins command a much higher price in the market. Perhaps the brown bear skins are as one in two to the black bear skins procured in the Territory. They are more common in the far North and Dr. Richardson in his Boat Journey overland in search of Sir John Franklin speaks of them, thus in page 218. "These brown bears are very powerful, and this same Gentleman (Mr. Bill) who told the above story informed me that one the Porcupine River, to west of the Peel he saw the foot-marks of a large one which having seized a Moose Deer in the River, had dragged it about a quarter of a Mile along the Sandy Bank and afterwards devoured it all, but part of the hind quarters. The bones were crushed and broken by the animal's teeth and from their size and hardness Mr. Bill Judged the Moose to have been upwards of a year old when it would weigh as much as an ox of the same age. The Species of these northern brown bears is yet undetermined. They greatly resemble the *Ursus arctos* of the old *world* continent, if they are not actually the same, and are stronger and more Carnivorus than the black bears *(Ursus Americanus)* which also frequents the McKenzie. The Grisly Bear *(Ursus firox)* reach the same latitudes but do not generally descend from the Mountains."

I forward a small case containing what I consider petrified *Enerinites* and shells of ancient days etc. etc. They will likely indicate the Geological period of our limestone formation, but I fear that will not be of much interest to you, who may probably be occupied by subjects of far greater importance.

There are many brine springs in the District to the West of our great Lake Basin, some of them very rich Yielding 33 per cent of salt. I may at some future period, if agreeable to you, send some of the salt for analisation and also some of our rocks and earths.

I will not forget your request respecting the *Coleopterous* insects of this vicinity, but I fear that I may not be able to preserve them right. I will also try to preserve some fish, and if I can some of our reptiles. Hoping that the three packages which I now forward by Mr. Charles Cavilier will reach you in good condition, and that on receipt of the same you will be pleased to drop me a few lines in acknowledgement thereof.

I remain, Dear Sir, Very truly your's,

DONALD GUNN, SERT

3. DONALD GUNN to SPENCER BAIRD

Novr 26th 1857 Red River Settlement

[SIA, RU 305, Reel 6]

[Gunn refers to a "Gentleman in Edinburgh" who was likely George Wilson, the director of the Royal Scottish Museum. The Museum was established in 1854, and Wilson contacted Governor George Simpson in order to solicit the assistance of the HBC in obtaining examples of North American technology. In 1856, Norman Wolfred Kittson (1814-88) openly opposed the HBC monopoly over trade in Rupert's Land when he began conducting business as a retailer in St. Boniface. He had been competing with the Company in the fur trade since establishing his own post at Pembina in 1843, but during the 1850s and 60s he expanded his challenge to include merchandising and transportation. When steamboat service was established between Red River and St. Paul's in 1858-59, Kittson played an important role in developing that service and in 1862, after almost twenty years of competing with the HBC, he became the company's shipping agent at both St. Paul's and Georgetown, Minnesota.]

My Dear Sir,

Your much esteemed favours of Septr 5th, Oct 3rd, and 10th came safely to hand, that of the first date by the Octr. Mails and the others by the Mail of the present month which came in on the 13th.

I am highly delighted in hearing that we have added something new to your Museum altho' the addition be nothing more than a few shrews. From its Indian name *Beaver Mouse,* I infer that the shrew from Nelson River is *sorex palustris* of Richardson. It is well known that the Beaver builds his residence along streams and marshes, and the shrews of the kind sent often take up their abode in a Beaver House or vault. The big Mouse is likely new to you in the states, for I have never met with any of the kind in this place. I think its range is likely from Norway House to the Bay. I have seen them in considerable numbers in dry sandy locations along the Bay. It or one very much like it changes its color, in winter, from brown to yellowish white. I regret that Mr. Isbister did not attempt any description of the habits of these Creatures. The shrew sent from here has been taken in the plains. I have not seen any of them in the Houses, altho we have taken one of the same kind last summer near the House. I am sorry that my short and imperfect acquaintance with the Creature will not admit of any attempt at describing its mode of life and No. of Young. I have fallen into error in stating to you that the *Winark* is not to be met with in this Vicinity, for this summer two of them were taken near this place, however, I think these are different from the former. The large Hawk (the goshawk) from Nelson River is the only Specie of that Genus which winters North of 49 parallel. The Eagles never remaining here later than the present date. They reappear in the latter end of March and are always the first of the migratory birds which make their appearance in our Northern regions.

There are two kinds of white partridges found along Hudson's Bay. The Partridge and what is called the Rock Grouse, the latter is not found in any great numbers so far south as York Factory in Lat. 57° but are plentiful along the sea coast at and near Churchill River Lat. 59°. I shall endeavour to get a few white Partridges in their summer dress. The Rabbits also in like manner. I cannot help thinking that the person who took the White Rabbit skin made a Mistake. The Badger skin procured here and a skunk skin were in the same Box with the Nelson River specimens, and all that I shall say on this subject is that whoever took them did not set a high value on his honesty. I did not send the Badger skin thinking it to be new to you, but to show that they are common here, but do not extend to the woody Districts on the East side of Lake Winnipeg. The skunk is not found at any distance to the N East of Lake Winnipeg. I sent several grey squirrels, two - perhaps three. We have many varieties of small Gulls in the summer season, in Lake Winnipeg, but likely there are more to be found along the Bay. There are no Gulls nor any other kind of Water fowl wintering in any Locality North of the 49° parallel. They all leave in October, the smaller ones earlier, and return in April when the Rapids are free of Ice. I shall endeavour to get as many Winter specimens of birds as I can, and I will endeavour to draw Mr. Isbister's attention to the small gulls and to the Wintering Birds of his neighborhood.

The Gyr Falcon does not Winter here. Is he a fishing hawk?

You say that you did not find the skins of Cat, fisher, beaver, marten, etc. as mentioned in Mr. Isbister's list. Mr. Isbister did not send the skin of any of these animals, except that of a Muskrat, which left here in the Nelson river Collection and which I fear has been taken to keep company with the Badger. Mr. Isbister is a servant of the Honourable Hudson's Bay Company and could not send these skins without purchasing them, which would at least cost five or six pounds Sterling. I understood Mr. Isbister is giving the dimensions of these animals whose skulls he sent.

I shall endeavour in future to be more particular in packing up. In yours of the 10th you acknowledge the receipt of mine of the 9th of September, via Toronto. I sent that letter by a friend who left here for Toronto on the 11th Septr *[torn]* thro Minnesota, but forgot to post the letter at St. Pauls Minnesota, but did so at Toronto. A Gentleman came to this settlement from Quebec via Minnesota in 14 days. I am sorry that I did not take the precaution to wrap the Geological specimens separately. Our limestone has not been much wrought or quarried, therefore we can get such specimens only as are on the surface and I have not yet fallen in with any teeth or bones.

Last year I wrote to three of my friends in the Company's service, one on the shores of the Bay and two in the far North. One of these Gentlemen did not return me an answer, the others stated that at same time they received my letter, they had orders given them by their Superiors to make collections in Natural history for a Gentleman in Edinburgh. One of them concludes by say. "I should have been happy to try and collect for you as far as I could, But as you very well know I am a Servant and must Obey those in power." Mr. Isbister resides on Nelson River about half-way from Lake Winnipeg to where that River pours its waters into Hudsons Bay. The eggs of Gulls may be procured in any quantity on the rocky Islets in Lake Winnipeg and along the shores of Hudsons Bay. Hawks and their Eggs are more difficult to be had.

In the month of September I had the pleasure of Mr. Kennicott's company for a few days. He gave us some useful lessons for collecting and preserving all sorts of Creatures. He also had the great kindness to leave me some alcohol and Arsenic. He has taken charge of the wolverine skeletons, one Flying squirrel, one cross Bill, all from Nelson River. From here one striped ground squirrel, one Striped wood squirrel, one Diver, a Grebe, and two or three little birds. The skins of two long tailed mice, one shrew and one mouse with a few Beetles in Alcohol, all of which I hope Mr. Kennicott will deliver to you on his arrival at Washington. I have since he left made a small collection of water Beetles, a dozen or so of our common Mice, two shrews different from those sent and a number of small fish. I have much pleasure in stating that I have received the very valuable present of Books which you had the kindness to send to the care of Mr. Kitson, who very kindly forwarded them by Mr. Cavallier who is passing the winter in this Settlement. You very kindly offer to send me more books and ask what size of packages could be taken from St. Paul to this place. Mr. Cavallier brought ship Bales that way last summer, but I think that small packages of one cubic foot each will be the most convenient and the most likely to find their way here, any of your literary newspapers can come by Mail and I have *Harper's Magazine* likewise. Bound books I believe are not taken by Mail to this place, but a few of such as you may have to spare could be sent to St. Paul's, and Mr. Kitson's trains could bring them and I would gladly pay costs. The reports of the patent office I think would be highly useful here.

You expressed a wish to know my "particular bussiness in life". I am at this present time managing a small farm, my own property, which is chiefly wrought by my family. Any books that you may send to me direct, as you did last the package sent last summer. The Newspapers or anything forwarded by mail please direct as your letters to me are directed. None of them has ever been lost.

A few days ago I called on Mr. Cavallier to inquire after the Alcohol and Thermometers, but neither of the two came to hand, nor will this winter. I have a Thermometer which I received by the ship of last season and I will as soon as I can command a little time, copy off my observations and send by mail to the patent office. You wished to know if I could afford time to enable me next Summer to go as far as Nelson River to collect Specimens. I think, If alive and in health, that I could take time but I fear the expense may prove an obstacle, the Journey could only be performed by water and for that purpose a canoe will be requisite and two men to Navigate it. I think the expenses of the trip would not be much below £25 or 20£, and I might not be able to collect anything that would compensate so great expense.

The expense of a letter from here to Pembina is 1d, and the same from Pembina to here. So you see that the expense I am put to is very triffling. You are desirous of knowing if any Trout be in our waters. There is none as far as I know In Red or in Lake Winnipeg. Hoping to hear from you by the first mail. I reman, Your obdt Servt,

DOND GUNN

N.B. the others sent by Mr. Kennicott was the same as the ones sent in summer

4. DONALD GUNN to SPENCER BAIRD

Red River Janr 11th 1858

[SIA, RU 305, Reel 7]

[In this letter Gunn expresses his opinion on the HBC's monopoly and on the political status of the Red River colony and the District of Assiniboia. During the 1850s, there was considerable agitation within the colony, as well as in Canada and Britain, for free trade, political and legal emancipation from Company interference, and bilingual services in the colony.]

Dear Sir,

My letter to you had been written before the Decr. Mail left but got mislaid so that I could not at the time, lay my hand on it and being, more than occupied with bussiness of a political character, which would not admit of delay. I could not command time to write another, and as I have found my original letter I send it by the Mail which is to leave on the 13th Inst. Much of my time has been occupied during the last year and is likely to so, for some time to come, in writing on the Hudsons Bay question. For we are determined to do all in our power to blot out the last remnant of the Despotic rule of the House of Stuart, and to have this country emancipated from the Iron bondage of the FenChurch Street Nabobs, to Unite it to Canada and have it opened up to civilization with all it's ameliorating blessings.

I have had a letter from Mr. Kennicott, date Nov 9th. I was happy to learn of his safe arrival at Chicago. And, as he wrote of going to Washington I have good reason to believe that he may be there at the time that this may reach you. If so, please present My kind regards to him, and at the same time beg of him to excuse me for not writing by the present opportunity and assure him, in my name, that I will not forget to write to him and on all the subjects which he mentioned, by the next Mail. The winter has been exceedingly fine, the lowest Indication of the Thermometer was 21-Journal registers of the Weather will be forth coming as soon as time will admit. Yours Sincerly,

DOND GUNN

5. BERNARD ROSS to JOSEPH HENRY

Fort Simpson: McKenzies River Dist. 28th November 1858

[SIA, HBC Corr Coll, Folder 36]

Bernard Rogan Ross (1827-74) posing for photo, circa 1865. Hudson's Bay Company Archives, Provincial Archives of Manitoba.

[George Gibbs (1815-1873) was a lawyer in New York before going west to the Washington Territory in the 1840s. He was lured west by the discovery of gold, but remained there twelve years during which time he served as botanist and geologist on the North West Boundary Commission. He was a Smithsonian "collaborator" in ethnology and philology throughout the 1860s, and he was also appointed secretary to the Hudson's Bay Claims Commission after returning to Washington.]

Sir,

I was honored last Summer by receiving from Sir George Simpson, a copy of your interesting letter, regarding the procuring of some Specimins of Natural History to complete your collection of North American Animals.

It gives me at all times much gratification to advance the cause of Science, as far as the limited sphere in which I move will permit: and in the present case the pleasure is doubly enhanced by affording any assistance to an Institution of such worldwide renown as the Smithsonian.

I have to inform you that a specimin of the Mountain Goat has been already procured from Peels River, and will be forwarded by our usual Summer craft. The animal has not got its winter coat: but I opine this to be of little moment. Should however the specimin prove, on receipt, not sufficiently good, I will, if you inform me thereof, endeavour to procure another.

Orders have been sent to the Posts where they are procurable for the Skins and Skeletons of the Raindeer, Muskox and whistling Marmot. We will also give directions to obtain if possible a barren ground Bear *(Ursus arctos Americanus* I presume).

There will be a considerable difficulty in procuring the Muskox and Bear, in a state complete enough for a museum; owing to the country which they frequent being so far out of the Indians usual range, and several years may elapse before I can procure specimins. If so I can assure you that it will proceed from no want of care nor attention on my part. In any way that I can be of use to your Institution you may command me.

As none of the Gentlemen of the North West Boundy. Commissn. are likely to visit McKenzies River, it will be out of my power to show them the attention that I otherwise would; but I am at present in communication with a member of that body, Mr. Gibbs, respecting the Chipewyan tribes of Indians, their offshoots and languages.

Any of your Publications, Pamphlets or Reports would be most interesting to me, and if sent addressed to the "care of the Hudson's Bay Company, Fenchurch St., London," would reach me in safety. I have the honor to be Sir, Your mo: obt Servt

BERNARD R. ROSS

Chief Trader, in charge of McKenzies River District

Address any communications to B.R. Ross Esq., McKenzies River Dist. Care Hudsons Bay Compy. agent, Pembina, Minisota Territory, United States

6. ROBERT KENNICOTT to SPENCER BAIRD

Norway House June 15th [*1859*]

[SIA, RU 7215, Box 13]

[Alexander Hunter Murray (c.1818-74) made his way to the Hudson's Bay Company, from Scotland, via the United States and the American Fur Company. He is best known for his role in founding Fort Yukon in 1847, but he was also important as the person in charge at Pembina during the 1830s, and at Lower Fort Garry during the 1860s. Donald Ross (c.1797-1852), originally from Stornoway in the Hebrides, joined the HBC in 1816. He served as Governor Simpson's confidential secretary in 1826-27, and he accompanied Simpson on his Columbia expedition in 1828. In 1829 he was made a Chief Trader, and in 1830 he was given charge of Norway House. He remained in charge of Norway House and the Norway House District until 1851, when he received furlough due to poor health. The Audubon Club was a small, informally organized group of Chicago naturalists. It was a predecessor of the more structured ornithological societies that first appeared during the 1870s, such as the Nuttall Club (1873), the Linnaean Society of New York (1878) and the Agassiz Association (1879). The Smithsonian Institution published an *Annual Report* which contained both business and scientific news. Articles describing original research were appended to the Executive Committee's financial report to the Board of Regents and to the progress reports on research and acquisitions. The *Report of the Commissioner of Patents (Agriculture)* referred to in this letter contains articles written by Kennicott on "The Quadrupeds of Illinois Beneficial and Injurious to the Farmer" between 1856 and 1858. Chief Factor Archibald McDonald (1790-1853) was also a member of George Simpson's expedition to the Pacific Coast in 1828, and while travelling west he recorded his observations on the climate, soils, resources, and potential for agricultural settlement in the northwest. He was later an Honourary Member of the London Botanical Society, and a contributor to the British Museum. Edward Martin Hopkins (1820-93) was Governor George Simpson's personal secretary and assistant. Hopkins, a native of Essex, joined the company in 1841 and spent thirty years at Lachine. He became a Chief Trader in 1847 and a Chief Factor in 1854.

In closing, Kennicott always sent regards to friends and colleagues, specifically the members of the Baird family and his comrades at the Smithsonian—the Megatheria. The Megatherium was the name given the extinct giant sloth of South America, and the adoption of this name for the Smithsonian fieldworkers' club conforms exactly to the process of "poetic popularization" identified within the nineteenth century sciences by S. Cannon (*Science in Culture*, 1978, p. 5). Indeed, Cannon's example of "poetic popularization" is the Megatheria. In one of Kennicott's letters to Baird, he identifies the "real" Megatheria as Philip Uhler, Carabus Ulke, Baron Ostensacken and the "shell men." This last group would have included the authors of the "Check-lists of the Shells of North America" (1860), namely, William Stimpson, Isaac Lea, Temple Prime and William Greene Binney. Mrs. Baird—Mary Helen Churchill Baird (1821-91)—was a well-educated woman from a Vermont family. She married Spencer Fullerton Baird in 1846 and their only child, Lucy Hunter Baird (1848-1913), was born two years later.]

Dear Prof Baird,

Hurrah! for Ft Simpson and musk oxen. I arrived here three days since and Sir Geo Simpson luckily for me came in the next day in advance of the boats.

He recieved me very kindly, as has every gentleman of the service that I have met.

Upon my stateing your interest in the fauna of the MacKenzie River region and my desire to visit Ft Simpson, he told me that though he would do all he could to facilitate my operations, with pleasure, he could not pass me *free* to the Mackenzie River, that not being allowed by rules of the council. But that, though it would be particularly inconveniant to send me this year, on account of the heavy loads and the fact that a clergyman and two clerks are going up, still he would direct that I should have a passage to Ft Simpson at the same rates that missionaries are charged if my funds were sufficient. Upon learning that I had still over 100£ he said that would cover the expense, and told me to make my choice between going to York Factory with a free pass or paying for the other trip. Before replying I begged that he would say if, for any reason, he preferred that I should not go to Ft Simpson. He said that as coming with such recommendation from Prof Henry as I did and he, understanding the interest the Smithsonian would take in the explorations of this region, he had no objections whatever to my going excepting only the trouble of adding to the loads, which he would not let stand in my way however. Having already made up my mind that this would be decidedly the best thing for the zoölogical operations, I told him I would prefer Ft Simpson. So I am to start in the boats tonight. In our interview Sir Geo. expressed much interest in my success and said he would write to Mr. Ross about me, as he has done in the most satisfactory manner. He told me that while at any of the companies posts, my living would cost nothing but that for provisions for the route, for my passage and for any special service on which I employed men I would pay myself.

At my request he gave me permission to remain two winters at Ft Simpson, If I wished. And if more funds, enough to enable me to work to advantage, can be sent me I will perhaps stay a second winter. As I learn from Mr. A.H. Murray, whom I met at Pembina and who has been at Ft Simpson and on the Yukon! that the whole of the Mackenzie River is a glorious place for birds nests.

I am to pay (I believe) 2£ per "piece" for my luggage, a piece being about 90 lbs. A passenger himself being reckoned at 5 pieces (you know I said I'd grow fat and heavy 450 is pretty good weight). I have about 4 pieces of luggage and three of provisions. Making in all, with myself, 12 pieces at 2£ to pay for. So the entire passage is only about $120, in addition to which I'll have to pay a cook but not much.

I cannot tell what my provisions will cost, nor can I get a bill, but will have my accounts sent to me at Ft Simpson this fall.

I have asked Mr. Ross, a gentleman at this post, to send you a copy of my account at this place. To this add about £10 to cover other orders and deduct from £142, and you'll see what I start with.

I will be as economical as is possible but shall of course be obliged to charge all expenses to the fund. I doubt my present supply of money being enough for two winters operations, while if more could be had it could be well used next spring. If money is sent me, it should be sent to Sir Geo Simpson at La Chine who will write to the north of its deposit and it will be placed to my account. I could get advances at any post if I were sure of having money to pay up. Did not some of the contributors say they would double their subscriptions if I staid over winter? And couldn't you poke up the Audubon club through Wilkins?

I'll write to him suggesting that they send at least the other $75.

Sir Geo Simpson just came into my room to say good bye and told me if there was anything whatever that he could do to assist me, to write him by the returning boats.

It would be very desirable that you or Prof Henry should write him, if you think proper, during the summer or winter, expressing your satisfaction at the facilities given me on account of the Institution. I'm not done work up here yet, if they will continue these facilities and I can get the means. Mr. Murry gives me a strong invitation to spend some time with him at Ft Pelly. In fact I'm treated as if welcome everywhere. I will probably find a large number of meteorological observers, and many collectors. I wish I had 50 copies of the instructions for collecting.

I suggested to Sir Geo the idea of getting up a museum at some point in the Territory, but the constant changes rendir this difficult and the idea was not taken up by him. But he has a private museum at La Chine where he has many fine things and he has written to Barney Ross this packet for musk ox and goat, and asked me to assist in preparing them if collected. I suggested several other points to which he should write, and asked if he would not direct that several be collected that you might get the duplicates. This he did. I said that I thought you might get him a cock of the plains, which grouse he has not got and particularly desired for his own collections. I told him I would ask you about it, and that you might perhaps spare some other western rarities. I did not specify not knowing whether youd like it, though I believe it would be a very excellant thing to offer a lot of birds for his private museum. And if you can do it I wish you would. The result to my operations would be desirable.

Can you not also put his name down for all the Smithsonian and Patent Office *reports?* Mr. Hopkins his Secretary expressed much pleasure at recieving from the Smithsonian, this winter, one of the *reports.*

Any pub docs whatever would be acceptable, I infer, from what Mr. Hopkins says and I told him I would ask that the Governors name be put down for the S.I. and Pat Office *Reps.* Anything more I of course did not like to propose sending till I knew your mind but Mr. Hopkins, if not Sir Geo, would take some interest in any scientific publication.

Sir Geo himself is—not meaning it disrespectfully but to speak earnestly—a *regular brick*! He is about 70, but active and energetic, not a bit old fogyish and very agreeable in his manners. Aside from his particular kindness and attention to me, I like him very much.

In fact all of these men are bricks, all intelligent and *gentlemen.* I hear Barny Ross is a splendid fellow.

You see I'm not exactly a martyr! to science when I go north. Frozen nozes, hard fare, etc. only will give zest to the whole.

N.B. A winter packet leaves Ft Garry for Ft Simpson about Dec 15th arriving at Ft S. early in March. By this any reasonable amount of letters may be sent for me, at least a couple of pounds they tell me here. And no doubt if Mr. Mactavish at Ft Garry (who starts the packet) was told it was important to you at the S.I., he would send several lbs.

So please send me at least one of the new insect circulars, and might send several of them and the others to Mr. Mactavish requesting that he put them in or not, according to the amt of room. If I could get a thousand or two good German insect pins I would like it, and they could be readily sent. Mactavish though very hard worked will, I infer from what I learn, take much pains to oblige you. He is really much interested in science, working daily at his microscope. Whether I stay two winters or not I ought to have a lot of arsenic, alcohol, etc. and particularly wide mouthed bottles with *good* velvet corks.

To send a piece (of 90 lbs) from Ft Garry to Ft Simpson will cost £2. But the things should be directed to me at *Portage La Loche*, the Height of Land this side of Lake Athabasca, where I would meet them on my return. Or if I staid at Ft Simpson, Mr. Ross, *who comes up every summer* with his boats to meet the Lake Winnepeg brigade at the grand Portage (P. la Loche), would take any packages down for me. And to here the tariff would be still lower. Anything sent might better be simply directed to B. Ross. He is in charge of the entire Mackenzie River district which includes the whole northwest, and does the duty of a chief factor. Thus anything sent to him personally would be well cared for.

Sir Geo told me I might have a couple or so pieces sent to me next summer by the boats on the same terms as now. And I've no doubt more could be sent if desired by asking Mr. Mactavish in time. Anything sent should be at Ft Garry before the 1st of next June to be safe, though the boats start any time from the 3rd to the 10th. As the company now send carts to St. Paul for their goods brought through the U.S., these pieces could perhaps be sent thus to Mr. Mactavish who could inform you how much could be taken. Could you not send me, a couple of collecting chests with about 12 gals of alcohol and 20 or 30 lbs of arsenic and some bugging bottles.

I will write again by the boats we will meet no time now.

The following gentlemen have promised to collect in zoology and make meteorological observations, and to them I wish you could send Pat office and S.I. *reports* with blanks and instructions.

A.H. Murry Esq Chief Trader Ft Pelly, Swan River
Mr. Archibald McDonald in package with above
Wm McMurry Esq Chief Trader Ft Alexander
Library of Norway House

These will be sent to their destination if placed in Mr. Mactavish's hands. Sir Geo says that any packages of books for gentlemen in the service will be sent from St. Paul if directed to the care of Messrs. Burbank, Hon Hud Bay Cos agents St. Paul.

I'll write by the boats. We are off now.

Kind regards to Mrs. Baird and Lucy and the rest please show this to the Megatheria.

Yours, for the North Pole!

R. KENNICOTT

7. BERNARD ROSS to JOSEPH HENRY

Portage La Loche, 25 July 1859

[SIA, HBC Corr Coll, Folder 36]

["Native superstitions" were occasionally cited as cause for poor collections. Kennicott, like Ross, noted native reluctance to sometimes collect or kill animals and he described how he overcame their hesitation by tricking them into thinking he had the prophetic powers of a medicine man, or by relieving minor aches and pains with medicines such as alum and morphine. ("The Journal of Robert Kennicott, 19 May 1859 - 11 Feb. 1862," in James, *The Scientific Exploration of Russian America*, 1942, pp. 95-97).]

Dear Sir,

I now forward, to the address of the Smithsonian Institution, 3 cases # 1.2.&3, which have all been carefully packed, and will, I hope, reach you in safety.

A List of Contents is now enclosed, and I am happy to state, that, with the exception of the Musk Ox, and smaller Siffleur or Marmot, the whole of your wants have been supplied. Great difficulty was experienced in procuring the Reindeer, from the superstition of the Natives. I have sent the Skeletons as well as the skins of the Goat and Reindeer. The Barren ground Bear is merely the skin, and is not a complete specimen: but is as good as I could procure at present: if it will not answer, I will endeavour to procure a better one as well as the Skeleton.

I have forwarded some Fossils, plants, organic remains, and stuffed birds, which may prove interesting and acceptable. Will you inform me in your next communication what animal the pieces of tusks and bones belong to? Anything in my power, respecting the collections of objects of Natural History, I will willingly do.

Address as before to "Care of Hudson's Bay Coy. Pembina, Minnesota Territory".

I remain, Dear Sir, Your obedt Servant,

B.R. ROSS

Ch. Trader in charge of McKR Dist.

8. BERNARD ROSS to SPENCER BAIRD

Portage La Loche 29 July 1859

[SIA, HBC Corr Coll, Folder 36]

[Andrew Murray (1812-78) was a Scottish naturalist and collector. He was a Fellow of the Royal Society (Edinburgh), President of the Edinburgh Botanical Society (1858), Secretary of the Royal Horticultural Society (1860) and a collector for the Royal Scottish Museum.]

My dear Sir

I received your very interesting favour of the 26 March a few days ago but from pressures of business am unable to reply to it at present as it deserves.

Mr. Kennicott has at last arrived here on his way to McKenzies River, where I shall be most happy to have him as my guest. The plan which I have drawn out for him is this fall to Liards, where he will be able to stop and collect some weeks, winter at Fort Simpson, Spring to Slave Lake for Eggs and aquatic Birds, Summer to either Resolution or Rae, descend in fall to the Youcon, pass a full year there, 3d year. The newly discovered Ross and Anderson Rivers, winter at Fort Simpson and go out summer 1862. I will I can assure you do all that lies in my power to husband his means, and his only cost while in my government will be merely his personal necessaries. Sir George Simpson has given him only leave for 2 years: but there can be no difficulty I should think in procuring an Extension of the period.

You will please accept my thanks for your most interesting works on the fauna of America. Every moment that I can spare I will devote to your Institution. I regret that I had not been in correspondence with You before, as I last year sent from Slave a specimen of the *Someteria v. nigra* to Mr. Murray of Conland—the receipt of which has not been acknowledged, tho I think it must be the specimen you mention. None of the natives had ever seen such a bird before, nor had I myself during a 6 year residence at that body of Water at Ft Resolution.

I send you 140 Tongues. Will You send 40 to President Buchanen from me. If I remember right I saw him in Derry—when a *very little* boy. I may be mistaken, but send him these at all events. They were cured at Fort Good Hope near the Arctic coast. You will divide the remainder between Yourself and Mr. Kennicott senior, and I trust you will find them good.

Any works on Ethnology will be acceptable, as I am at present engaged in the study of these Northern languages, which I speak passibly.

I must now conclude being much hurried but I will write at length by Winter Express.

I remain, Most truly yours

B.R. ROSS, CHIEF TRADER

9. ROBERT KENNICOTT to SPENCER BAIRD

Ft Simpson Mackenzies River, Nov 17th 1859

[SIA, RU 7215, Box 13]

[One of Kennicott's lengthier letters, several individuals in it need identifying. Professor Turner may have been W. W. Turner, an ethnologist who edited a French manuscript of Chinook jargon that was published in a Smithsonian publication entitled *Vocabulary of the jargon or trade language of Oregon* (1853) by Dr. B. Rush Mitchell (*SIAR*, 1863, p. 26; 1868, p. 436.) Carl Robert Romanovich von der Osten Sacken (1828-1906) was a member of the Russian diplomatic service in North America. He lived in Washington from 1862 to 1871, where he was a member of the Smithsonian scientific community and a noted entomologist. His work on *Diptera* was especially important, and catalogues compiled by him were published by the Smithsonian Institution in 1858 and 1878. Dr. Edmund Andrews (1824-1904) was a military surgeon during the Civil War, but he practiced medicine in Chicago and taught at Chicago Medical College. He also taught at

Rush Medical College, and at the University of Michigan. John Woodworth was one of Kennicott's Chicago friends. He went to Washington for a three month period in 1861 to help Baird process incoming specimens. Carabus Ulke was one of Kennicott's Smithsonian friends and a member of the Megatherium Club. He was also a photographer. He took the picture of James Lockhart included in this volume. Dr. Helmuth may have been William Tod Helmuth (1833-1902), one of the founders of the Homeopathic Medical College in St. Louis, Missouri. Dr. Helmuth was a resident of Pennsylvania until 1858, when he moved to St. Louis. Louis Agassiz (1807-73) was one of North America's most prominent scientists during the nineteenth century. He moved to the United States, from France, in 1846, and he became Professor of Natural History at Harvard University in 1847. He was a popular lecturer on zoology and palaeontology, and he was an outspoken defendant of controversial theories such as polygenesis and recapitulation. He opposed both Darwinian and Lamarckian theories, preferring a more traditional teleological explanation for the origins and development of organic life. He became a Regent of the Smithsonian Institution in 1863. Dr. Henry Philip Tappan (1805-81) was an ordained clergyman and an educator. He taught at New York University during the 1830s, and he was the first President of the University of Michigan. He also founded an astronomical observatory at Detroit. Professor Alexander Winchell (1824-91) was one of Baird's field collectors and correspondents during the 1850s. He collected botanical and geological specimens in Alabama while employed there as a teacher. He became a Professor of geology, zoology, and botany at the University of Michigan in 1855, and he directed the Michigan geological surveys, 1859-61 and 1869-71. Henry John Moberly (b. 1835) signed on with the HBC in Lachine in 1854. He twice resigned from the company, but following his final reinstatement in 1870, he founded Fort McMurray.]

Dear Prof Baird,

I have been said to be *constitutionally* opposed to labor. That was my brother Charlie's mild way of saying I was infernally lazy always. Added to this, I am now in the laziest community I ever knew. With this preface, I dare not offer it as an excuse, I proceed to confess that I have, during the last month, been most shamefully idle. I could have prepared quite an elaborate essay on this country; I have done nothing of the kind and my notes made hastily on the voyage remain unaltered. So much for my promise to send you my journal. I have also neglected as yet to get any considerable information in Ethnology and shall probably send little or nothing to Prof. Turner. Tho a number of gentlemen at the different posts have promised to send you lists of words etc., and you may recieve something by this packet.

Mr. Ross expressed a desire to take charge of the meteorology and he will send in his observations made thus far. Observations will also be made at other posts. One thermometer I sent to the Yukon, another to Ft Good Hope, the third to Ft Rae Slave Lake. The other three were left for Michipicoton and Ft William, Lake Superior, and Ft Pelly, Swan River, near the Saskatchewan. I have distributed meteorological blanks to all the posts in the District, and at those we passed on the voyage up.

Since leaving Portage La Loche I have collected several hundred specimens of plants.

At my more legitimate work, zoology, I can say I have not been lazy at least. At this only I've worked well; and indeed it has required constant hunting to get anything. What I have succeeded in obtaining you see by the register. What birds and mammals I've observed and not obtained, you'll find in the short notes I send.

Since September I have often hunted all day without seeing a specimen except whiskey Johns and pine squirrels, and lately red polls. But there are yet fine things to be had this winter beyond what I've already obtained. As soon as the packet goes off, which it will the first of December, I shall begin more diligent search for small quadrupeds and the few winter birds. And I hope to trap some martins, foxes, carcajous, etc., though I confess I shall trap as much for the purpose of keeping my energies awake (if it is possible) as for the sake of the specimens and fun of the thing. For it takes a great deal of time with small returns, this trapping. Mr. Ross has very kindly promised to let me keep all the fur animals I catch myself, unless Sir Geo Simpson's orders are to the contrary. As for purchasing any fur animals from the Indians, that is impracticable as I should be obliged to pay the London prices of the furs, viz about 23 shillings sterling for martins and other things unless I get Sir Georges permission (if he even can grant it) to buy them! in proportion. As I very stupidly neglected to have Sir Geo Simpson arrange these matters at Norway House, I am as yet ignorant as to what I can expect to get of such species as have valuable furs.

Mr. Ross has however been so kind as to promise that he will himself endeavor to send you a lot of the larger species as carcajous, sheep goats, musk ox, reindeer, etc. Tho' of the latter four species I believe there will be no objections to my taking specimens—if I can get 'em, which isn't so easy to do! However at Ft Liard I have the promise of mountain sheep and carcajous. At Ft Halcott Mr. Ross thinks he'll get goats, at Ft Norman, Ft Rae and at a post on Athabasca, he has offered a very large price for musk oxen and at La Pierres House and Peels Riv I am almost sure of obtaining goats deer etc.

A party of Esquimaux from the coast were brought up to Ft Simpson this fall. Two men, a boy, a girl, and a woman (The girl remains here this winter). They were infinately more intelligent than any Indians I have ever seen; after a single lesson they could all skin birds and mammals very well and a number of the specimens I send were skinned by them here. I gave them several pairs of scissors and a couple of knives besides presents of tobacco pipes, etc. and had them take skins of a bird and mammal lest they should forget any part of the "modus operandi." They expressed much delight, and repeatedly promised to collect me specimens of each kind of bird and mammal on the coast as well as eggs; as they are honest and so intelligent, and withal so much pleased with their reception here, I do not doubt they will keep their promises; the more so as they will expect good pay; and the value of several dollars would equal their greatest expectations. So I'm well pleased with the prospect in that direction.

We arrived here about the middle of August; for a week or two I collected but little, as during that time all the gentlemen of the district, with one exception of one, were here. After that I hunted like a good fellow and flatter myself I've gotten several things that will call one of those *pleased* smiles onto your face, that I'd as willingly see as find a new insect; perhaps I couldn't honestly say snake or bird. You won't laugh at the three toed woodpeckers I send *now* I wot, nor at the titmice with brown heads and *too* much color on the breast, nor at the rose breasted grosbeaks that are rosy backed too, the dozens of fine deer mice that have pouches in their cheeks, and the *arvicolae*, especially some that look as if each side of their noses had been scorched to a reddish brown! I do wish I could just see you a moment when you opened the little package I shall send you with this. But alas I fear you'll expect more than I have yet obtained. The egging was a dead failure; but wait. Next spring! In the latter part of September I went up to Ft Liard in a boat that was returning from here, and came down again with two savages in a small canoe. I was too late to get marmots of which there is a small species (size of *Sper. franklinii*) and another as large as a manx. But these Mr. Mackenzie the gentleman in charge of the post will get next summer. And while at the Fort I taught a boy and woman to skin birds and mammals well, as I also did a couple of Indians from the *mountains* who promise lots of strange things next summer and among others a white tailed ptarmigan! and what I suppose must be the little chief hare.

A low spur of the Rocky mountains passes close to Ft Liard and extends to within 25 miles of Ft Simpson. My martin road ends within a few miles of these mountains.

Nothing pleased me better in my trip to Ft Liard than getting *arvicola xanthognatha*. I could only get specimens by shooting them, and found them only in one place. I spent several hours going up, and as many more coming down, in getting what I did. Digging them out was impossible.

On my return from Ft Liard, after three weeks absence, I found several new birds arrived as you'll see by the register, but the cold weather coming on nearly all soon left. Then when I ought to have begun writing I got a lazy fit that nothing would break. I would sit several hours pen in hand without finishing a page. As a last resort I took a savage, a couple of blankets, provisions (tea, salt and "old Indian" i.e. dry meat) and axe, and went off to make my martin traps. Made a road 20 miles long with over a hundred traps (Dead-falls baited with fish heads), besides snares for cats and rabbits. I was out six days, temperature always below zero and part of the time 20° below; I had no choice but work hard or freeze; a carcajou kindly followed me and pulled down every trap for over two thirds of the distance so that I had the fun of putting up such as I had cached baits enough for. Of course I was very wrathy; and finally came home and have succeeded in working since. Now I know how to cure my fits of laziness I'll do it promptly hereafter, or *freeze* in the attempt. But do you know? 'tis not so easy freezing here! Fancy how it is. During this time I was out I was not more warmly dressed, excepting my feet and hands, than I have been at home on a chilly autumn day, and at night I slept under two blankets only for several hours in a temperature below zero after our camp fire had gone out! Is it not strange when you consider, too, that I have been thought peculiarly sensitive to cold? You may suppose the "back fat"—the fat, which is often very thick from the moose's back—formed a considerable portion of my diet. I could eat blubber like an Esquimo.

I need not tell you that I am unusually strong and healthy. Rather better this than getting the yellow fever or at best suffering from malaria in Mississippi.

These are my plans for the future supposing I have a supply of preservitives from home and a little more of *the needful.*

This is not the best point for collecting that is certain. For birds eggs 'tis no good. I find *no old nests* excepting a few robins nests and there is no place for most water fowl. I have sent a box containing such things as I can spare to Big Island at the foot of Slave Lake. Next March I'll go with the packet to Big Island, and Mr. Hardisty will send his train of dogs for me to that place and I'll cross Slave Lake to Ft Resolution, his post. He is Chief Trader, tho' under Mr. Ross, and will he says be glad to have me come. If I can get a train of dogs I shall buy them—cost with man to drive perhaps $25. If I cannot get them I'll walk to Big Island. Ft Resolution is decidedly the best point for birds and eggs and as good, at least, as any other for other things, and I do not hesitate to adopt this course therefore. Perhaps instead of staying there till the boats return from Portage La Loche I will after the egging season, which is June and July, go around the northeast side of the Lake to Ft Rae and thence to Big Island where I'll wait for the boats going down to Ft Simpson in August. This will depend on my means, and the cost of Indians and canoe to take me.

Next fall I will go down with the boat to Peels River cross the mountains and down the Yukon to Ft Yukon, remain there till the succeeding summer in August, then with the boat go up to La Pierres House, stay there shooting goats! during autumn and early winter; then before it gets too cold cross the mountains to Peel's River where I'll spend the third winter, making in company with the gentleman in change an excursion to the Esquimaux country near the coast provided I have means to purchase dogs. Before the river breaks in spring, I'll walk on snowshoes to Ft Good Hope and from there a boat goes to Portage La Loche. Then if you think proper, instead of working about Lake Winnipeg and Red River till fall, (and you get Sir Geo's permission) I'll go down to York factory with the same boats that take me from port La Loche, stay there four or *[five?]* days and be back at Red River in September. Or if you say so, and Sir Geo Simpson gives me four winters here, I will on returning from the Yukon spend a summer in going to Anderson River and spend the fourth winter say at Athabasca, staying there till Mr. Ross passes about the 20th of July. Now I call that a brilliant programme!

But if I can stay only two winters I still say Hurrah for the Yukon! For I'll go down next fall all the same, arrive there probably about a week before hard frosts; so will have time to get fish, snails and even plants and insects, as well as birds and mammals; then go up to La Pierres House, before it gets very cold, (Do it *comfortably* with dogs in twelve days) and go out as proposed, unless I stay at Fort Yukon and go down to the Russians the next summer. As I have said, it is not impossible, though hardly probable without much expense ($100 or $200), that I could from the Yukon reach the Russian posts in summer. The Indians of the upper Yukon are afraid of the Russians and of the Indians below. But had I the necessary letters written in Russian I could very likely get an Indian for an enormous bribe ($25 would purchase this *enormous* bribe in goods) to go down to the nearest post and return with a couple of guides, then I could take a Ft Yukon Indian with us to secure the safty of the party from the Indians of the upper Yukon and very easily float down with all my *traps* in perfect safty. I don't know whether it would be difficult for you to get letters from such Russian authorities as would be respected by the servants of the Russian American Fur Co., but if you can do it with Baron Osten-Sacken's aid please try even though as I say, the chance of my success is small. That is if you think proper under all the circumstances. You must also understand that Sir Geo Simpson may not desire that anyone should go down the Yukon. It is not yet desired that there should be any communication with the HB Cos posts, or that the Russians should know the country! This is suggested by Mr. Ross, who thinks it may not be for the interest of the H.B. Fur Co to have the Yukon explored and known. But as my objects are purely scientific, and I'm quite willing to give my word that I will publish nothing and give no information to the Russians bearing on the fur trade, I think Sir Geo might not object.

Since Mr. Ross has, just now, given me the above information respecting Sir Geo's *probable* objections to such an expedition, I have modified my plan to this extent—that I shall only ask permission to go down the Yukon in the event that I can only stay two winters in this District; for if I ask to go at any rate, maybe he won't let me go to the Yukon at all. And at the same time I believe you will very likely prefer me to go back through H.B. Tr. in case I can spend more than two winters.

To return to the letters—Letters sent to Red River in time for the packet leaving Dec 1st 1860 would reach me on the Yukon in June 1861. These could at that time only be made available in case you had written me by the packet this winter or the boats next summer, that I might remain at Ft Yukon relying on their coming. As I wrote you something on the subject from Port La Loche I'm of course in hopes to recieve the letters, or news of their *not* being forthcoming before I start from here next fall. In case they are sent would not Baron Osten Saken be kind enough to give me English copies of the whole and write me a letter stating that I am at the H.B. Cos post, Ft Yukon, waiting for means to reach the nearest Russian post, and that I desire that if possible a guide or two be sent to me with whom I can descend the river in a canoe which I'll purchase at Ft Yukon.

Though I have said so much about this plan, I repeat there is but little chance that I can carry it out even if you desire me to. Sir Geo. may not wish Russian Indians to learn too well the way up the river, and worst of all the Indians of the lower Yukon have long been at war with those of the Upper Yukon and their fear of being *spiflicated* is the main cause of there being no communication. But where there is a will there is a way.

But I have neglected to give in my programme another possible plan for the fourth summer and winter. This is it; go in the spring (early in June) from here either to the Anderson River as proposed (or to Ft Halcott *in the Rocky mountains*), spend the summer and get back here in time to go with the boat to Ft Resolution the last of August. There hire two savages to take me up to Ft Chipewayan on Athabasca Lake in time to catch the boat for Dunvegan; go in it up the Peace River to Dunvegan far to the south west in the plains on the slope of the Rocky mountains, winter there; and in spring a party carrying leather always goes across the Rocky Mountains to Frazers River whence I'd easily reach a steamboat, or I could in spring return from Dunvegan leaving May 1st, and reach Athabasca in time to get to Norway H. in the Athabasca brigade July 1st.

I am often inclined to think that the fourth summer might possibly be better spent at Ft Halcott in any event than in going to Anderson River; unless it should happen that by that time I can reach the post to be built on Anderson Riv before the birds arrive. To do which I must go with dogs. Perhaps by then there will be a road across from Peels Riv. So long as I remain in the District, which includes everything north of Athabasca, Mr. Ross can send me where he chooses so that I can make my arrangements on the spot with him.

Now comes the grave question, will I have money to carry out any or all of these plans? I enclose an estimate of my expenses and proposed orders and expenditures till next summer. I have not yet recieved my bill from Norway House so cannot say what I owe there, but I suppose it is about £40 including charges for my journey up. You observe in my estimate I include nothing for charges as passenger out; Mr. Ross says he shall not charge me anything and I don't suppose I'll have to pay for myself or baggage going home. If I have no considerable expenses beyond what I now foresee, I could get through, by economy, to the fall of 1860; and be as near home as St. Paul, with £50 ($250) in addition to the £42 I estimate I'll have left next March. But on the one hand it may possibly become highly advisable that I should get in my own allowance of flour, tea, etc. for the last year and I have left little or no margin for unforeseen expenses; while on the other hand Sir Geo may have directed that an extra allowance of provisions be sent for me next summer, in which event the provisions mentioned in the estimate will not be sent, probably.

I think I mentioned to you the fact that on arriving at Portage La Loche I found that no extra allowance had been sent in for either Mr. Kirkby the missionary or myself, both guests at Ft Simpson. As Mr. Ross and all other gentlemen have only a certain yearly allowance, and that of course no larger than they require, I am thus compelled to live upon Mr. Ross; though he tells me that if the allowances sent next summer are complete he will not be short. You must understand, however, that, as Sir Geo did not know before seeing me that I was going to Ft Simpson, it was then too late to send in an extra allowance for me, even had he thought of it, as the boats were overloaded and some pieces were actually left at Norway House. I am inclined to think, however, that he couldn't have remembered the need of it; for not only was no allowance sent for Mr. Kirkby, but in talking with Sir Geo. I asked him if while here I would not cause the officer in charge extra expense? He answered "No. You will be living at the expense of the company and you're welcome to eat all you can." or something to that effect. I did not know at the time that only the officer's regular allowence was sent in so could not percieve any need of a special allowence for myself.

Mr. Ross has written to Sir George on the subject from Portage La Loche, and he says he has very little doubt that enough will be sent in for me. Still to be on the safe side, I order from Norway House part of a clerk's allowence of tea, sugar and flour, as it would never do to sponge on Mr. Lockhart at the Yukon; he having only a couple of bags of flour and two kegs sugar, with a scanty allowance of tea, rice, rasins and butter. If the extra allowences ordered by Mr. Ross are sent in, my order will probably not come but even if it does I shall need it for voyaging from place to place or when away from the posts. A full clerks allowence for a year costs the company (at prices charged the officers) about 20 or 25£ at Ft Simpson, including cost of getting it in. Should you have plenty of money, for this expedition, collected, it might be advisable to offer to pay for this unless you learn that an allowance is sent to me. The fact is I'm sorely puzzled what to think. Mr. Ross says he thinks the allowance for me will almost certainly be sent, and that if it is not it will probably be only because there is no room for it in the boats. In the latter event of course my order can't come. Then too it is possible that Sir Geo. might have intended to send the allowence and would feel offended at an offer to send it in to me at our expense. I shall not mention the subject to Sir George myself (I'll send you a copy of my letter), and in writing my order to Mr. Sinclair at Norway House I'll say I order this lest Mr. Ross should not have enough to supply me with provisions in case I went to the Yukon. You may very likely know already, or may hear from Sir Geo Simpson as soon as he gets this packet, what will be done in the matter. And I must leave it to you to write to him or not on the subject as you find advisable. But if you write at all before hearing from me by the Athabasca brigade next July or August (when I'll tell you, having learned by the winter packet, his answer to Mr. Ross), do it immediately upon recieving this, and if you think your letter may not reach him at La Chine before he starts for Norway House last of April or early in May better send a duplicate letter to him at Norway House, so that he'll get it before the boats leave there for the North. You observe, however, that there is no actual necessity for you to make any further arrangements for next summer as my order will be sufficient in any event. I only mention all this about your writing at once because if you have plenty of funds belonging to the expedition, an *offer* to bear this expense of my allowence, which as I say, is worth to the company some £20 or £25. It might make a material difference in what Sir George would do for me in reference to staying and travelling about. This is in point of fact the greatest actual expense I make the company, the provisions being worth as much as money at Norway House.

But even if the allowence has been sent they could hardly feel offended at your writing on the subject. You'll judge better than I what is proper to be done. You of course understand that I by no means complain of my living. I only fear lest I should by my presence deprive the officers of part of their few luxuries.

You can tell Sir George if you should write to him that I'm quite delighted with my life here, and my chief fear is that I may not be able to stay long enough.

To return to the question of funds — supposing I don't have to pay for any more provisions than I've already ordered and you send me in preservitives for specimens I could probably, as I said, get through even with $250 more than I have now. But this would be hard work and place me in a decidedly disagreeable position very frequently, besides forcing me to operate on a limited scale Still send me this and tell me to do it and I'll try it. At any rate I shall stay next winter. I hope to get more next spring on Slave Lake than I've gotten in all yet. And I've no intention of losing so much time for nothing. I can't afford to go home next spring.

I hope however that somebody will embrace this magnificent opportunity to do "*grand chose*" for science and come down liberally to the tune of 500$ or 1000$. Then I could get the good things easily. For you must know that there is very little chance of my ever killing such things as musk oxen, barren ground bear and Reindeer, goats, etc. myself, and I can only hope to get them by hiring the Indians to bring them in from a great distance. You see by the bill I enclose that my personal expenses here will not amount to very much.

I have already explained in my last letter why I ordered so much wine and brandy. For myself I would not have cared for it and this I shall give to Mr. Ross and to Mr. Lockhart whose guest I'll be at the Yukon. Nothing would please them more. Therefore I trust you'll not think I've done wrong. Mr. Ross says he will give you a strong hint to send with the books he is going to ask you to buy for him, some whiskey. And pray don't medicate quite all the alcohol—but most of it!

I send a list of articles wanted for the third winter. I of course know that you may be unable to send all I ask for. But you also may be able to send more; and remember that nothing will be wasted here. Now I have to buy goods for paying the Indians, and a few knick knacks that would cost a very trifle would give even better satisfaction than anything I could buy here.

I have written home to have some "goodies" and alcohol or whiskey sent to Mr. Ross. With these are to be packed some articles for me. Mr. Ross gets them in free.

Please send me a large quantity of *good* strychnine for poisoning animals. But get that with large crystals. I bought a bottle in Chicago and have fed about half of it to the carcajou who pulled down my martin traps. He appears to have found it an excellent *tonic.* At last Mr. Ross gave me some good strychnine, I put it out and yesterday brought in my Carcajou who had died within a rod of the spot where he found the bait.

Please write to Wm. Sinclair, Esq Chief Factor Norway House stating which of the boxes you send are the most important, lest they can't all come up. But if Sir George chooses he can have them all sent at any rate. I trust you have so gotten into the good graces of Sir George and Chief Factor Hopkins, his secretary, that they'll do all they can to advance your plans.

I of course anticipate the *possibility* of recieving neither money nor collecting material from home next summer; but in that event I shall still go to the Yukon! If that most liberal and honourable University you wot of, were to pay me what they owe I should work on that for a time, though I'd prefer to let my paternal have it.

My gun was but a poor thing. But I'm in luck; last fall I exchanged it for a small London gun no 24 gauge that a few years since cost £15. It is in good order except that the stock is cracked. But twill stand several years. The gun is very light and throws small shot beautifully. I killed with it, with ball, a wild cat. The first shot at a very respectable distance. Balls, B.B. shot, and *poor* powder I can get in any quantity here, but very little small shot, tho I still have say 3 lbs no 10 and 15 lbs no 3 shot. I have also over 10 lbs of H.B. powder left brought from Norway House. You must understand however that sale shops here are as unlike as possible what they are elsewhere. The effort is to sell as *little* as *possible,* as it doesn't pay expenses. But as servants here are paid the same as at Norway House, they must sell their goods at the same price too. The amount that a person may take, even an officer, of any article is quite limited. Though Mr. Ross has very kindly let me have all the goods I want of such few things as are still remaining. Of course you'll not suppose I've myself used any considerable proportion of the good charged to me. Goods from the shop are the only legal tender here.

My stock of clothing, blankets, etc. is now quite full and I hope I'll have to buy very little more. But what I do get the Expedition must pay for! Though for that matter I hope the 25$ I've myself furnished will very nearly cover the whole.

As I'm to stay so long I might as well refer to my collections at home. You, Dr. Andrews and John Woodworth must arrange what is to be done in case the Chicago academy can take proper care of the museum. I shall be *rather* short of cash by the time I get home and I don't want to give up my collections entirely. But I prefer to leave the matter to yourself and I give you the veto power over any measure you don't think proper. If a carcajou should spiflicate me or I get frozen on a trapping excursion this winter then I hereby bequeath my collections to the Chicago acad. provided they can take proper care of them. I trust the aforesaid mighty institution will raise £25 or £50 *at least* for my expedition. But I'm not going to beg very hard for assistance from any quarter, at least I'm not going to promise too much. I know very well that what I'll get will certainly be worth all twill cost. And I'm not actually under obligation to collect, only for those who have already contributed money. But in all arrangements of this kind I'm represented by you, who will I know do whatever is right. I wonder if some of your European correspondents wouldn't contribute. Nothing except a *very large* contribution will be acceptable from Agassiz tho! Prof. Agassiz doubtless would for I shall certainly get lots of fish though I've gotten but few yet. 3 species salmon ("Red" *inniti* and "black") on the Yukon, white fish, pike, suckers, etc. plenty. I have also a few fossils (shells) and on the Yukon you know are wonderful vertebrate fossils. Prof. Leidy had better do the liberal. I've gotten a few good insects at Ft Simpson especially beetles. Tell Carabus Ulke I've plenty of what I suppose is that larger species resembling the *something or other ruscarius* that Dr. Helmuth had one of. He'll know what I mean, if I don't send him one. (which I have done)

About June 20th the boats start from here for the Portage La Loche, a boat having previously gone down to Ft Good Hope and returned. The ice breaks up late in May in the river. I cannot tell yet about the arrival of the birds, with any accuracy. Mr. Ross says the geese arrive at Ft Resolution in April though the ice in the Lake does not break up till late in June, and sometimes even till July.

Are Indian dresses and impliments wanted? That is, I mean, shall I get any large number. A suit of Indian dress complete would cost about £1 perhaps less. Small models less, and of some kinds still less.

I suppose for some museums such thing would be desirable. I'm getting a few things of the kind.

What shall I do about mammal skeletons.

Please keep my father advised as to what arrangements you make with Gov Simpson as he will then know what I am to do etc. Can you not after a time send him this letter, I give you some details that I may have omitted in writing home.

If you send in any books please send some to the Norway House library and get into Mr. Sinclairs good graces if you can. He really decides about loading the boats and can see that my pieces all come in if he chooses. Though Sir Geo Simpson also can settle that matter. You must understand that there is danger that all my pieces cannot come, but ask Mr. Sinclair if he cannot bribe one of the steersmen to take an extra piece or two if necessary. The men sometimes take a few pieces extra recieving the pay themselves ($10 or $12 per piece).

Mr. Moberly formerly of Jasper House about whom I wrote is not to return and the "minutes of council" don't say where he goes or who is to take his place.

I got no letters from the United States by the fall packet so have not heard from home since I left.

I wish you would get Carabus Ulke to send me two copies of your photograph — one for Mr. Ross. I've written to him myself about it.

Do you want to know how we live here? There are at the post Mr. Ross, Mr. Onion a clerk, Mr. Kirky an Episcopal clergyman (of Church Missionary Society) nicknamed "Brother Tadger" (see Mr. Wellers friend in Pickwick papers) and myself, for gentlemen. There are also about 8 or 10 workmen half of 'em with half breed and Indian wives. The men are two Norwegians, three orkneymen and the rest Red River French half breeds. Several Indians are still remaining also at the fort. The "Fort" consists of five square houses arranged in a hollow square open in front and facing the river. The buildings are very good. The "gentlemens house" has a large kitchen behind and is divided into a mess room, a bedroom for Onion and I each, a library and bedroom for Mr. Ross. These are on the ground floor. Above are summer rooms used by the gentlemen in fall. About a cord of wood is burned in the mess room daily.

We live well; our bill of fare consisting of dry meat (tongues finished), potatoes, fish, tea and an allowence of sugar, bread and butter with puddings of rice or rasins on Sundays and Wednesdays! In fact we have quite enough. At the other posts below they get no potatoes and the allowences of other *imported* articles is small. They raise some barley here which is ground in hand mills and most of it given to the men for a summer voyaging allowence, the rest being made into a villianous liquid they call "beer". You see what efforts are made to procure the element of conduction!! Mr. Ross this fall used part of a keg of sugar in the attempt to make wine of wild black currants; but tis worse than the beer, even. Luckily I'd poisoned all my alcohol and

Rum! The gentlemen are by the regulations not permitted to import liquor though they manage to get in a little.

We breakfast (now) at 8 or 9 o.c. and have dinner at 4. Card playing, until we began writing for the packet, has been the regular employment in the evening. Though hereafter I mean to insist on being permitted to write at least part of the time.

The officers duty is almost nothing beyond his actual presence. A little less than two months in the year is sufficient for all the writing. No wonder then they become lazy. Mr. Ross and Brother Tadger are the only industrious men I've seen here.

I find that bird and mammal skins dry very well here without arsenic, in winter at least, and I find some "arseniced" skins break very easily. I'm not sure but it would be better to prepare them in winter without it. I shall try some with the alum and saltpetre. On rabbit and martin skins I put no preservitive except about the legs and head. The rabbit skin sent had no preservitive upon it I think, but on the ermine I put alum and saltpetre.

At length I have succeeded in getting an ermine. It seems to be *P. richardsonii.* Is it not queer that the sharp tailed grouse is here. I didn't get it but had a good and near view of several and cannot be mistaken in the species. I left my spec. of *Bonasa umbellus* at Ft Liard so cannot compare it with the description. It will come down in time to go out etc. I'll doubtless get more. I've seen no ptarmigans yet but they must be here now or will come soon.

About the notes I send, they are perhaps not worth publishing in their present imperfect state but do as you like. If published please add the names of the species I do not identify. Next summer I'll send from Slave Lake a much more elaborate and complete set of notes and then you'll have the specimens for corroboration. Whenever I have given dates I have seen the birds myself and unless well known, shot and identified specimens. And I think I've not erred to any extent in giving species on the authority of others.

My bird and mammal book I've sent on with other things to Slave Lake and am using Mr. Ross' here.

The species of which I send specimens now are marked in the accompanying Register with a +, but not the same numbers as I do it from memory. Nor am I sure I've sent the number of each written above the + though about it. By the spring packet I'll send more if I get anything new. I suppose the *Lepus* from Portage La Loche is the same as this. I tried in vain this fall to get them in summer pilage here. They are exceedingly shy and cannot be shot. There is but one species here.

I hope you'll get specimens of various kinds next summer from nearly every post in the District. Mr. Ross has written to every one and I supplied pamphlets of instructions and a little alcohol to each and arsenic to some, besides giving some lessons in the fall on bird skinning. Mr. Ross upon my telling him that not he could write would prove uninteresting to you tore up his first letter and wrote you a second rather lengthy one.

As you are so fond of gossip! I may tell you Mr. Ross is to be married next summer probably; his intended meeting him at the Portage. You also like descriptions of people. Mr. Ross is about 32 years old, large, very strong and pretty corpulent, fair haired and "complected" as the Hoosiers say—the top of his head bald. He is rather good looking. He is a little choloric, though he prides himself on possessing the contrary temperament. He is certainly very intelligent and well read considering his opertunities as he entered the service at 15 years of age. Like your humble servent he has a very good opinion of himself and is in fact *rather* conceited. Never forgives any one for proving him to be in the wrong, and indeed will never admit that he is in the wrong if it is proved. He likes flattery very well though he would repel such a statement with extreme indignation. He is an inveterate practical joker and never looses an opertunity to chaff anyone most unmercifully. Still on the whole I like him decidedly as a man, if not as a companion.

Excepting Mr. Ross, Onion, Jones and Mr. Clarke all the gentlemen of the district are scotchmen or half breeds. Lockhart who is in charge of the Yukon I like best of the whole and I'm sure we will have jolly times together. He expresses himself delighted at the prospect of my being with him.

If I've sent anything new from this region please name a species after Mr. Ross.

I shall try to send next summer some dried skins of any large fishes that I cannot get young of. I've been unlucky about getting fishes. I tried in vain for them in the Liard and was nearly frozen in the attempt. I have only succeed in getting a few on a bat here in the Mackenzie. But there are plenty here if I can get at them.

My watch has stopped and I send it to you in the package of specimens. Will you not hand it to a watchmaker, and if it is worth repairing have it done and sent to me by mail in time for the boats. The expedition expenses must include this or I won't make meteorological observations!

Apropos, I shall after this month make the meteorological observations myself here. Though Flett the postmaster will also keep a register. Mr. Ross is tired of it. In the registers which Mr. Ross sends now he has made allowence for the inaccuracies of his thermometer.

I send a flying squirrel skin from Ft Liard which is not numbered. All the insects and snails are from Ft Simpson.

I have written letters to only the following (besides those to folk at home and officers at H.H.B. Co.)—Prof. Henry, Baron OstenSacken, Ulke, Mr. J.E. Wilkins of Chicago, Dr. Tappan, Dr. Kirtland and Dr. Andrews. I've only written a short letter giving my plans to Sacken and Ulke, of course I've told all the necessity of writing to me or yourself instan, or if they would be in time. I mentioned the Russian letters to Sacken. I've suggested *mildly* to Dr. Tappan and Mr. Wilkins the idea of further contributions from the club and University and more strongly to Dr. Andrews and Woodworth respecting the same from Chicago University and academy. I don't think I've been any *too* modest in representing to these gentlemen what grand things I am bound to do and what a devoted *martyr* to science I am. I have most courageously asserted that I shall cheerfully permit the Yukon Indians to spiflicate me, if need be; knowing, as I will, that my specimens already collected on the Yukon will have been safely sent home before by way of Ft Simpson! Now ought not that to move 'em to the disbursing point?

Speaking of finances, if you have plenty of funds belonging to the expedition please write, in your letter to Chief Factor Wm. Sinclair of Norway House, to have my order for sugar, flour and tea increased to a *full* clerks allowence including, however, 24 instead of 15 lbs of tea as I cannot get more here as the others can. That is do this unless you know an allowence is to be furnished me. I send for as little as will possibly do me, not daring to expend more, and I don't want them to think that is all I would require for an allowence either! But if I have some of the *element of conduction,* ie Brandy, Lockhart will for the sake of that, share his other allowences with me and I should if sugar were used in cooking learn to "prefer my tea without it," though when no grease is to be had in cold weather one needs sugar the more for its carbon. A third bag of flour would be the most acceptable *addition* to the allowence besides the tea already mentioned. Tea is worth, and will sell for, ten shillings to a pound sterling per lb here tho it costs less than two shillings at Norway House; so it is always useful if not wanted for my own drinking. But you *Grahamites* below little know with what gusto tea, tobacco and *fat* are consumed at the north.

You observe, however, that it is not because I would not willingly myself do without every luxury, if necessary, that I say so much about provisions. But to live at a post, a parasite upon the officer in charge when his supplies are actually insufficient for himself alone, is the only thing I'll *not* endure in order to get to the Yukon.

I trust you have John Woodworth under your wing. How Does Prof Winchell get on with his survey. Wonder if "Wilberforce" will get the Illinois survey. Woodworth and Dr. Andrews with my materials added to their own, would do it better.

Mr. Ross sends Lucy a pair of Esquimaux earrings! Fear she'll find them large for wearing. I'll have her some *rare* buttons next spring.

I've not time to write to Stimpson and other Megatheria now but will in the winter.

Kindest regards to Mrs. Baird, Lucy, the General and Mrs. Churchill.

Yours always

R. KENNICOTT

10. BERNARD ROSS to SPENCER BAIRD

Fort Simpson McKenzies River, 26 Nov. 1859
[SIA, HBC Corr Coll, Folder 37]

[Ross refers to his library in this letter. His collection forms the bulk of the Mackenzie River Fur Trade Library housed at the Hudson's Bay Company Archives. There are numerous works of history, geography, literature, and theology in addition to the many scientific monographs found in Ross's collection. The H.B.C.A. also possesses eleven of the twelve volumes of the *Reports of Explorations and Surveys to Ascertain the most Practicable and Economical Route for a Railroad from the Mississippi River to the Pacific Ocean [Pacific Railroad Survey Reports]* (1855-60). This series consists of catalogues of the birds, mammals, reptiles, and fishes collected by the army, in addition to the geographical information, maps, and travel accounts associated with the surveys that were sponsored by the Secretary of War.]

Dear Sir,

I must, previous to the departure of our usual winter Express, endeavour to answer, though inadequately, your very kind and interesting letters of the 26 March and 16 April last.

Through the instrumentality of my Officers and my own exertions, the various wantages for the Smithsonian Institution will be supplied I confidently expect in a few years. The Musk ox will be the most difficult to obtain, but time and perseverence will I expect overcome all the obstacles. I have already got a few fossils collected, from the Anderson River—a few Specimens of animals and some Esquimaux curiosities, and manufactures. If things of the latter discription would prove acceptable I could furnish a considerable quantity. I have written to the gentlemen in charge of Posts throughout the whole of this and Athabasca Districts to make collections for me, and I trust that what will be amassed previous to the opening of the Navigation will be something of interest. This will be entirely independent of Mr. Kennicotts labours. They will be collected at my expense, and I hope that the Smithsonian Institution will gratify me by accepting them as a gift. The Eggs I will pay particular attention to, though I fear that it will be the work of years to obtain a complete collection.

Fort Simpson is rather unfortunately situated, as regards making a good collection. Its situation, at the junction of the turbid Liards with the McKenzies River, prevents many fish being in the waters. Wild fowl having no feeding places on the River, or suitable lake to resort to in the vicinity, pass on directly. At a distance from the barren-grounds, none of the animals frequenting that tract of country ever come here: and the mountains are at too great a distance to draw Specimins from them. In other things a tolerable variety of Species may be got; Small birds, hawks, owls, mice, squirrels, wolverines, Martens, minks, Foxes (*fulv.*) Black Bears, Beaver, and Insects are tolerably numerous. Some of the birds, pigeons for instance, coming here as to their northn. limits. In the spring—Backs grayling—*Salmo McKenzie*, and white fish may be caught in small quantities, in a little river falling into the McKenzie opposite the Fort. In this River water hens, Sawbills, and Buffle-headed ducks are not uncommon, while solitary Harlequins are stock, teal, widgeon, Pintails, Canvass backs, Black ducks, Golden eyes, and other ducks, besides plover of various kinds can be collected. As there is a considerable farm here growing barley and potatoes with tolerable success, this is the best place I have seen for insects, and I made a very considerable collection here one Summer for Sir John Richardson. In the bird *line,* thrushes, blackbirds, woodpeckers, pigeons, Kingfishers, whiskey Jacks and Ravens are numerous. The Rook does not come north of Big Island, where it is rare—at Fort Resolution they are very numerous. Mice are very plentiful, as also are squirrels. The Fur animals inhabiting this tract of country are Foxes, (Red, Cross, and Silver) wolves, wolverines, Bears (black and brown) Beaver, Lynx, Martens, Minks, Muskrats, otters. Fishers are nearly unknown, though one or two have been taken about Big Island, and at the debouchure of Slave River into Slave Lake they are plentiful and I have caught as many as 10 in a Season there with steel traps or gins. Some of them equal in size to an ordinary otter, and, one *very old fellow* weighed 18 lbs. These animals are very strong and mischievous, often doing as much damage to a line of traps as a wolverine. Their colour varies, with age, from an almost black tint to a rusty grey. For the purpose of your instituting comparisons of the geographical distribution of the Fur-bearing animals in this vast District I enclose you a Statement of the quantities of the various kinds of peltries received at all the posts throughout this District for *the last 10 years*, and also the general or total returns of the District for *16*. This will give you an idea of their relative numbers far better than I can in writing upon the Subject. I wish you to consider the details of this paper AS STRICTLY CONFIDENTIAL, it is what I would not communicate to anyone but a thorough man of Science. The publishing of the details would seriously compromise me; but the inferences of a Scientific nature drawn from it cannot be objected to. Might I turn your attention to the remarkable circle of increase and decrease that each decade exhibits. In nearly all the Fur-bearing animals this is observable, but particularly so in the Martens—the highest years in the decade 1845-55 being the extremes and the lowest 1849, nearly the central one. Migration is the only reason that I can assign for such a fluctuation in their numbers; but if so where do they go to? We have posts at the most Northernmost post that they could subsist at. I have not at hand materials to form an opinion by observing if they increase and decrease simultaneously all over the Territory, both on this and the other side of the height of land. We never find them dead as we do the hares, which are quite as regular in their periods of appearance and disappearance. The latter in the winter of 1848 I found dead at Dunvegan, Peace River, in their forms, on repeated occasions. Such were evidently diseased there were boils over their bodies, principally on the neck and foreshoulders, and inside the stomach of some were found species of hairy, foetus looking substances.

It is extremely probable that your supposition, of the birds that penetrate furthest north being the largest, as well as all migratory animals, is a correct one. The more weakly ones will naturally remain along the road; and the same thing is I think observable respecting fish, regarding the distance that they come up rivers from the Sea. I will with pleasure aid in obtaining a Series of Measurements to determine the matter. Of the *resident* Northern animals being larger I am doubtful.

Your enquiry respecting the *Frontalis* I will endeavour to reply to at a future period, when I have investigated the subject at more leisure and with more scrutiny than I have heretofore done. My experience heretofore has been only that of a Sportsman, though of this kind I have had an ample supply during my 13 years residence in this district and 17 in the country generally. It strikes one that I have seen often birds among the flocks of Whitefaced Geese, answering to your description. I am also of opinion that an additional species of the *Berniculi Canadensis* will be discovered, of a size intermediate between the large kind and the Hutchins Goose. In the Hpls. or Snow Goose there will be noted probably 3 species, instead of only one as now. The largest kind, which arrives very late in the Season, in fact only passing Slave Lake when the leaves are formed on the trees, has a reddish yellow head and neck, much like a Swan. Another kind is not quite so large: and the head is either pure white or with very little yellow on it. The third, and rarest species, is in appearance similar to the latter; but is much smaller, being about the size of a large Mallard. The first Goose seen at Resolution is generally a solitary, large Canada about the 18 of April. In a few days, or a longer period, according to the advancement of the Season, more of his brethren arrive, seldom in large flocks—parties of from 5 to 20 being about the common size. Next the *Hutchins* and *Albifrons* appear in very large bands sometimes containing hundreds, and the two kinds often flying intermixed. The Snow Geese come latest, in immense bands, and the Yellow headed species last, and they are always the fattest—their size approaches that of the large Canada Goose. These remarks all apply to full grown Spring birds. To the movements, habits, food, incubation, and sizes of waterfowl (gulls excepted) I have paid much attention. For many nights, during the shooting season have I sat watching their motions from my "decoy" hut, made of twigs, and grass, and some thousands at least of the different kinds must have been shot by me during my residence, as until lately I have been a keen and successful sportsman, killing as many as 30 geese, besides ducks, gulls, plover, etc. and Sometimes a crane or Swan also in the same day. The white and red Crane are common at Fort Resolution. Two species of Swans, the Trumpeter and American are plentiful. Ducks of all Kinds, bittern, gulls, terns, Pelicans, eagles, hawks, owls, mice, shrews, small birds, fish, insects are very numerous, and eggs particularly Gulls can be obtained in any quantity. Altogether it is the best place for a general collection of any in this District. It was here that in spring 1858 I shot a specimin of the *Somateria v nigra,* and which is I think the same that you mention in your letter to me as having been shot at Slave Lake. I forwarded it to a Mr. Murray of Edinbro', but have not as yet received an answer to the note which I sent accompanying it. I am sorry that I had not been in correspondence with you as you would certainly have got the specimen. Mr. Murray I know nothing of, further than from a circular that was sent through the country, and his omission to acknowledge the receipt of this bird, will be certainly the best way to prevent me from taking the trouble of any sending him any more. I scarcely think that I will be able to procure another. It is not a fresh water bird and from its leaness when killed had evidently been driven inland by *stress of weather.* From the Anderson River, when we get a post established there it may probably be procured through the medium of the Coast Esquimaux. A complete Series of the aquatic birds of this District will be very interesting, and we must endeavour to procure a complete set, any duplicate of these that you do not happen to require, I would wish to be kept for me in case I should wish after leaving this country to form a collection, and I think that I can promise you a general collection of Specimins from these wilds, such as has never yet been seen in any collection in the world.

I am at present turning my attention to the Ethnology of this District in the fullest extent of the term; and I am now collecting my data. Vocabularies will be sent me from all the Posts. Dresses, manufactures, habits, customs, superstitions, I will minutely investigate. A series of measurements of stature, etc. are in progress, and I have sent for a large and complete Photographic apparatus to enable me to deleneate the Indians and the Scenery, as well as the Animals of the country. This will be my Study for several years to come, with the *stock* language I am well enough acquainted to talk fluently on ordinary Subjects, and pronounce correctly. The character of the Indians I am well up to, and I wish if possible to write a history of the discovery, a geographical discription, and a sketch of the Natural history of McKenzies River and Atha. Districts. The latter will of course be mere compilation, as I cannot nor do I pretend to the character of a Scientific naturalist, though on nearly all the subjects and branches I have a smattering more or less. Of the various instruments of reflection I am a tolerable master, indeed I passed the two or three first years I spent in this District in *amusements* with the Sextant principally. I was appointed in the Summer of 1850 to survey and map the Youcon and Liards and Pelly Rivers, and to determine the situation of our posts by a series of astronomical observations. Circumstances of a public nature prevented my journey, for which I am very sorry, as, although a very poor proficient in the various branches of Natural Science I am certainly superior to any of the officers in here. I am fond of reading and possess, for this remote place, quite a Library. It numbers about 500 vols. and is of a very catholic order. Almost a little of everything in it. It is most deficient in works of Science, though I have amongst others Cuvier, Todds *Materia Medica,* Carpenters *Zoology,* Kirby and Spence, Humbolt *Cosmos and superposition of Rocks*, Roberts *Conchology,* Admirality *Manuel of Scientific Enquiry*, Castle on *Poisons*, besides having sent this year for Agassiz *Compar. Physiology*, Lyells *Geol. Works*, Richardsons *Geology*, Kirby on the instincts of animals, Elliottsons *Human Physiology.* Would you therefore be kind enough to expend the enclosed draft in the purchase of such works in the various branches as you think would be useful to me.

George Barnston (1800-83) in 1877. Notman Photographic Archives 45, 131-B II, McCord Museum of Canadian History.

Some of the other volumes of the *Pacific railroad report* I would like. : the General Statistical one with views and the Ethnological one. Any lists or vocabularies of the Indian languages West of the Rocky Mountains, in Russian America or Siberia, would be most acceptable as without these I will not be able to trace the analogies between the tribes inhabiting this part of the world and those races. As far as we know the Russian Indians, they are of the same stock as the Chipewyans. Any instructions of how to proceed in my investigations will be of great Service, as will in fact anything touching on this branch. If you could also insert a copper collecting can containing about 2 Gallons of good whiskey into the centre of the case to dilute the reading, you would confer a favour and I will return it full of Specimins. This can be sent under the *supposition* that it is for collecting Specimens, for as this is a *Temperance* district, I have to do the thing "sub rosâ." For Security the things should be packed in soldered tin cases within wooden ones, and the *gross weight* of a package should not exceed 100 lbs.

I have seen neither Snakes nor toads since my residence in this District. The Slave Lake Indians *have a name for them*, though there are none in the vicinity of Fort Resolution, nor have I heard of their ever killing them. I have enquired of Mr. and Mrs. Flett, the latter a native of Peels River, if Snakes or toads were seen either there or on the Youcon, and they both state that such is never the case. I am therefore of opinion that Sir John Richardson must be mistaken, but I will make further enquiries upon the Subject.

Every assistance in my power will be given to forward the objects connected with Mr. Kennicotts mission, and I am sure that the other officers of the District will be also willing to do so. He will have free passages throughout the District to whatever post he wishes to proceed to, and freight outwards for such packages as he may transmit to you.

Packed along with the Specimens which Mr. Kennicott is now sending you is a pair of Esquimaux ear-rings, which I wish you to present to your little daughter, of whom Mr. Kennicott speaks repeatedly, from me. They were given me by a fair lady of that nation last Autumn.

I trust to be able to furnish some material for you in the way of notes on the habits of the fur bearing animals and water fowl before the Spring. Meantime with best Wishes

I remain dear Sir, faithfully Yours

BERNARD R. ROSS, C.T.

11. GEORGE BARNSTON to SPENCER BAIRD

Michipicton 15 Mar. 1860

[SIA, HBC Corr Coll, Folder 2]

[James Stewart Clouston (c.1823-74), son of the HBC recruiting agent in the Orkney Islands, joined the company in 1842 as an apprentice clerk but worked as a labourer until 1855. He left the company in 1859, but returned in 1861, becoming a Chief Factor in 1864. Livingston's *Travels* refers to David Livingston, *Missionary travels and research-*

es in South Africa: including a sketch of sixteen years' residence in the interior of Africa, and a journey from the Cape of Good Hope to Loanda on the West Coast; thence across the continent, down the river Zambesi, to the eastern ocean (1857).]

My Dear Sir,

I have at last, after much talk and difficulty, succeeded in procuring a full grown Male Caribou of the woods District. Your "rangifer Caribou". He was shot some ten miles to the northwest of this Establishment. For two days I have been superintending the skinning, and the disjointing, boiling and scraping of the bones an operation performed with adroitness by an old Halfbreed hunter, a Young Indian and a couple of women. I regret having had so little proper preservative for the skin, yet I think that the mixture of salt and pepper, with a little alum and saltpeter and a pinch of camphor will take it safe to Washington where a proper unction may be applied. The cavities of the Cranium and spine I have plugged up with tobacco. Unfortunately it being so late in the Season, the Horns had been shed, and again so early in the spring, the fresh ones have not sprouted, but I made the black membrane under which the new antlers are developed to be very carefully kept entire with the Skin, and the slits of these membranes can be neatly sewed. They were made by the knives of mischievous children, before I got the animal. In one respect the want of the horns and the appearance of the surface whence they sprout is interesting as exhibiting the state of the skull or head just as the new horn is germinating. I am told that the young and vigorous males have their sprouts already out, but that this old Buck having had a severe hard winter, is in that respect in a less forward state. This accords with the late casting of the hair of weak or sickly animals. You will find the hair of the chest somewhat worn by the snow Crust. The Idea of a celebrated Comparative Anatomist that the front palmations were providentially given to the Caribou to enable it to perform the awkward process of "shovelling" away the snow from its food in winter, was quite a mistaken one. It uses its fore hooves like the Horse for that. At the season when the snow is deepest, if it depended on its horns, it would perish, for *then it has none*. In the woody region the Caribou in winter depends much on the Lichens and mosses that attach to the scrubby Branches and the stems of Trees, although these are not its favorite food. The Habits of the Caribou are not much Known in this Country although its flesh is the *finest* venison that can be eaten. With cranberry Sauce it is the great dish of Hudson's Bay. I shall send you very full measurements of this superb Buck, taken immediately after the body was brought to me. If you can procure from Greenland or the north, the Skull and bones of a male of the *R. arcticus*, you will then be able to ascertain the anatomical differences. On the plains to the northward of Churchill, towards Seal River, I followed these Deer with the assistance of Chipewyans. The natives looked upon them as distinct, and they are smaller and go in much more numerous bands, not mixing with the wood Deer, although they approach the edges of the wooded Country in the winter. We were in a Tract half occupied by Clumps of stunted Fir and Juniper, or rather Larch, the two last names being synonomous amongst our Orkney Servants, and half by open ground covered by Lichens and Moss, the hollows and banks of Rivers, yielding Arctic willows and plants.

Possibly the *Rangifer* met with on the uppermost waters of the north Branch of the Columbia, as well as near the upper waters of Frasers River, may be different from the Caribou, but I am disposed to think it the same. In coming up the Columbia in April of 1831, Mr. McLeod who was of our party, killed 3 or 4 out of a band of 6 or 7. We were then two days march from the Boat encampment the great mountain forks of the northern Branch. I believe it will yet be found that this Deer is the same as the Caribou, and that it ranges along the Rocky Mountains from 51° of North Latitude by the sources of Frasers River and Finlay's Fork of the Unjigah, as far as the streams that penetrate the Main ridge to empty themselves into the MacKenzie. Here there is quite a connection of wooded country eastward to the MacKenzie, and although the Peace River or Unjigah Valley is essentially a Moose Deer Tract, Yet towards lesser Slave Lake on the South of that and all along the north bank of the Athabasca River there is a broad Belt of Vegetation fit for the Caribou. In the Lesser Slave Lake District, The Moose, *Rangifer Caribou*, and *Cervus Canadensis* may be all met with, the first and last Keeping to the poplar Ground and small openings, the Caribou to the less fertile Fir and Larch Land. The Moose, by the bye, ought to have been Called the American Elk, not so the Canada Stag. An Ignorant person long ago made the Mistake, which now causes a wrong impression and is apt to create confusion as to allied species.

The Gentlemen at Lacloche on Lake Huron to whom I wrote to procure me a Moose, replies that the animal has been extirpated in that quarter, so that I am sorry I shall not be able at present to procure You one. On the outskirts of Canada the Moose is still to be found, but decreases in numbers. If you have correspondents on the borders of Maine or New Brunswick they may perhaps procure one. Again on the North Side of the St. Lawrence in the region lying between Lake St. Johns and Grand Lac and Temiscamingue, and especially that portion of it watered by the upper streams of the St. Maurice, there are still a considerable number of Moose, and it is from these that the Iroquois and Algonquin Tribes derive their Leather, which they prepare so beautifully for winter Shoes. About forty Years ago, to the westward of Lake Superior City there were a few Moose, but probably they have been rooted out ere this time.

A good old thermometer which has been giving us regular note of the temperature since 1st of January — was unfortunately broken the other day, by a Young Gentleman having taken it from a screened position to another more exposed to wind, where he might observe it more easily. Have the kindness to send me one of your best and include the expence with the rest of the Charge for Map, Books and other things which you may be pleased to make. I shall settle with You through Clouston as last Year. I expect You will send me preserving Liquid or Soap along with the wide Mouthed vials and a few Shrew Traps. If Livingstones *Travels in Africa* can be had for 4 or 5 dollars, please enclose them in the Box, and five or six fathoms of fine wire Gauze to enclose fry and small fish in any streams for the purpose of observing their growth etc. a wdt 3 feet broad x 20 long will answer. This ought not to be very expensive. I have much to write and must conclude hastily with respects to professor Henry and the best of Good wishes.

Yours sincerely,

GEORGE BARNSTON

12. ROBERT KENNICOTT to SPENCER BAIRD

Fort Simpson Mackenzies River H.B.T. March 23rd 1860

[SIA, RU 7215, Box 13]

[Alexander-Antonin Taché (1823-94) was a Roman Catholic priest sent out from Montreal in 1845 to establish Oblate missions in the northwest. In 1853 Taché became bishop of St. Boniface and under his auspices missionization advanced northward into the Mackenzie River and Athabasca District. John Lawrence Leconte (1825-83) was educated as a physician but independent means allowed him to pursue his interest in entomology. He was an expert on *Coleoptera*, although he wrote on many topics in natural history. In this letter, Kennicott extends greetings to other acquaintances at home. Mrs. Churchill—Lucy Hunter (d. 1862)—was Mrs. Baird's mother. Her father, General Sylvester Churchill (1783-1862) had spent his life in the military, and had served in both the War of 1812 and the Mexican War. He retired in 1856, but volunteered his services to the Union government after the Civil War broke out. He also used his position in the military to assist his son-in-law's collecting program. He supported Baird's suggestion that military personnel be outfitted and instructed in conformity with a program being developed at the Smithsonian to expedite scientific field collections.]

Dear Prof Baird,

Last night the longed for "Packet" from Red River arrived, and at about the usual time. In it I get your letters of May 6th, May 22nd, June 3rd, July 17th, July 22nd, August 29th and Oct 12th. The last enclosing lancet and gun nipples, and containing my latest news from you. It was in answer to my letter of July 6th.

I very much regret that I could get no reply to my letter from Portage La Loche. Yet your forethought has rendered it of less importance.

My two great questions are answered, viz., do you think the long stay here will be paid for by the small returns I can make. And, secondly, can money enough be raised, without too much exertion on your part, to keep me going. When taken into consideration the prospect of a winter and summer at the Yukon (after my summer on Slave Lake) and my *present* prospect of a winter and part of a spring, at least, on the Anderson among the musk oxen and Eskimos, I feel sure your recommendations to stay will be stronger when you know of these.

I have no further advices from the Governor, nor has he answered Mr. Ross' letters from the Portage, so I can't tell what he will say to my staying longer than two years. But his consent gained, and my life spared, I will very probably at the moment you read this be crossing the Rocky Mts near the Arctic circle or floating down the Yukon, to remain a year (nearly) at Ft Yukon.

I wrote you by the winter packet my plans for three winters operations (including this winter); but as another party of hungry savages may have eaten the packet met, as they once did, I'll repeat what I wrote with such modifications as I have since made.

If I don't get Sir Georges permission to remain longer than two winters I shall go to the Yukon next fall all the same, and walk from there to Ft Good Hope the following spring. That is unless I can get out by way of Russian posts as proposed, when I'll spend the bird nesting season at Ft Yucon and go down the River about June or July. But if, as I hope, Sir George gives me a third winter I'll stay at the Yukon till August 1861, then go in the boat to La Pierres House spend some time there and at Ft McPherson on Peels River just across the mountains, and, recieve my letters and outfit at Peels River by the boat from Ft Simpson on the 5th to 10th of September. Then I'll make a trip (about 10 days walking) on the first snow to the new post on the Anderson River in the Barren Grounds north of Bear Lake and only a degree and a half (about) from the Arctic coast. At this post (which will then be fully established) I'll winter and probably can stay till open water and yet be in time to catch the boat going from Ft Good Hope to Portage La Loche. But I am very much inclined to like the Idea of spending the summer (till August) of /62 on the Anderson and then coming to Ft Simpson to meet the fall boats from the portage in August. Then to go with the boat to Ft Resolution early in September, hire a canoe and go on to Ft Chepewayan Lake Athabasca, and thence with the companys boat up to Dunvegan on slope of Rocky Mts. Or if I can't go to Dunvegan push on to Isle a la crosse by open water (or even further) and walk to St Paul on snow shoes!!

I sent you an estimate of the cost of two more winters here, that is one on the Yukon and one at Peels River. To spend the third winter on the Anderson instead of at Peels Riv would cost very little more, if any difference at all. *It could be done for about $500* if Sir George sends me in allowences. The cost of a years allowence (of tea, flour etc.) if paid by us and freight charged would be somewhere in the neighborhood of £20 if I remember right; but I sent you an estimate. But if I pay Indians for specimens to any extent I ought to have more.

I still have, I think, over 40 pounds sterling on hand after paying cost of trip to Slave Lake. My expenses this winter have been very light. My bill at Norway House was a little over £42 but includes £1 paid servent and for trip to Portage La Loche.

Now you say you can raise $250 or $300 a year. This covers my estimate even without the £40 on hand, so I feel safe for two winters more at least (observe that I include an order for a partial allowence from Norway House, for next winter, in estimate of what is *already expended*, so that is one winters allowence covered.)

Still, I doubt not, that the fact of my going to the Yukon and Anderson will make a difference in the liberality of your contributors. And I've already mentioned that while I'm not going to let pecuniary difficulties stop me if possibly to be over come by my economy and privations, yet I should on the other hand not only *prefer living comfortably* but believe it to be the best policy for the interests of the expedition that I should have the means to do so. The value of money in obtaining specimens it would be superfluous to speak of. The officers kindly permit me to take what I want from the stores so I can always offer the Indians good bribes. But they are lazy superstitious Devils and I've got nothing much from them yet.

I start for Slave Lake in two days. A *good* train of 4 dogs has been sent for me so I can carry pretty much all of my collecting materials. I shall take about 100 lbs besides provisions. I have a pressing invitation from Mr. Clarke to go to Ft Rae, his post, and I think I shall send on nearly all my things from Big Island to Ft Resolution, while I will go to Ft Rae, stay a couple of weeks or so and then go to Ft Resolution. This I can do with perfect safety any time before the middle of April. Thus I will get some things at Ft Rae and teach the folk there to skin birds etc. My only hope of getting ptarmigans this spring is by going to Ft Rae; but I fear they will even be gone from there ere I arrive. As Mr. Clarke was so kind as to send his dogs for me I'm in hopes that this trip to Slave Lake will cost me very little. I may possibly get barren ground carabou at Ft. Rae.

Possibly however I'll now go on directly to Ft Resolution and make a trip to Ft Rae in July in a canoe. But I can't fully decide till I get to Big Island.

I have been *hybernating* this winter. I've been in the most apathetic state you can imagine. I am in perfectly good bodily health, in fact never before in a long time was so long without being in the least sick. But my mind as I say has been in a state of hybernation. I look forward with pleasure to my long walk to Ft Resolution or Rae. But to write has been *purgatory* to me. In fact I'm mentally, but not physically, villianously lazy—if you can make any sense of that. Very unfortunately there have been no ptarmigans here this winter and even rabbits are scarce. My fur hunting has been a dead failure. And I can't trap even mice till lately when there has been a little warm weather.

I've had neither *pain nor pleasure* this winter except when I had mal de racquet as mentioned further on. The time has been *[page torn]*erfully short to me and I look back on the winter pretty much as if I'd been in a long sleep.

An Express arrived here from Ft Liard just before Christmas and by it I recieved an invitation from the officer at that post to visit him. As I was hunting here in vain even for whiskey Jacks, I though I wouldn't lose anything at least by going and that perhaps the journey might awaken my dorment energies, so I started with the party on its return (January 3rd). I had a sled and three small dogs, too weak to carry more than their load so I could only ride about four miles each day. We made the trip of 260 miles in a little over seven days. It was my first real journey on snowshoes and I got male de racquet the first day. This is a *horribly* painful inflammation at the ankle of the muscles which lift the anterior part of the foot. Lucky it was then, that I was apathetic and with my sensibilities dulled or I believe I should have given up in despair and delayed the party, as it was I only felt the bodily pain and kept limping on. I didn't exactly limp either for one leg was as bad as the other. Two days of the voyage the temperature at Ft Simpson was about 40° minus (the coldest weather of the winter here and at Ft Liard). I froze my feet, hands and face a little but this, fatigue, and blistered feet were as nothing compared to the infernal mal de racquet. However I got to Ft Liard at last and you may be sure didn't take long walks for a week after. I spent about six weeks at that post and a short time since returned walking all the way back excepting a few miles the first day. We came in nine days one of our party knocking up and delaying us. But I got no mal de racquet and was not much tired either. Expect to go to Slave Lake in fine style, and have some hopes of making a decent voyageur yet. By the way Dr. Rae is said to have been the best snow shoe walker ever known in these parts—beating the natives.

With the return of warm weather and the reawakening of nature I hope to find my wits again.

Writing as I've said has been purgatory to me this winter, and I've not yet gotten anything into a sufficiently readable form for sending home. In fact have not yet written fifty pages of my notes, and the little done only covers my journey as far as Norway House. Probably I shall not send home anything at all for publication. But just write rough notes and mark them up after I go back. Maybe I'll keep awake next winter at the Yukon however. I send nothing this time.

I console myself however by thinking that I didn't come to write about the country, but to investigate its zoölogy and at this I can always work with interest and *enthusiasm.* In fact I only awaken from this infernal apathy when I get hold of some interesting animal. But I shall *keep trying* to write about the country.

I have lately found an *arvicola* resembling A. *gapperi.* I send specimens, and I send a shrew by this packet. By the winter packet I sent you a number of specimens to be forwarded from Red River by mail.

I recieved a letter from Mr. Mactavish written early in December. He is a brick. Promises to send me all the things I ordered and to help all he can—tells me to "command him in any way."

I hope you will be able to send perservitives in time for the boats next summer, but if you don't I shall just do the best I can with what I have left. I shall keep over a little arsenic however and Mactavish will send the rum and he says will send some *arsenical soap*, he has, if he can find it.

But if I am alive *I'm bound* to go to the Yukon. THAT IS SO! specimens can be kept without preservative if need be!

I shall write more fully by the portage boats and I have reason to expect that letters will reach you earlier next fall than last, and in fact you may get them very soon after this.

Mr. McFarlane, the officer now in charge of Ft Good Hope and who is to establish the post on the Anderson walked up from Ft Good Hope arriving here a few days since—made the trip in 15 days (about 500 miles). He says he went during the winter to the Esquimaux at the *mouth* of the Anderson. He took a thermometer, one I had given him, mercury *constantly frozen* for 22 days of his voyage! Yet he says they only got frozen *one* day that they were short of provisions and obliged to push on with a heavy wind in their faces.

I have mentioned my being out with the temperature 41° below zero and one day the wind blew very sharply but fortunately from behind. I was *not as much* wrapped up as I would have been on a very cold day in Illinois excepting perhaps my feet and hands, yet I didn't suffer very much from cold! This winter has been unusually mild—41° below—the coldest at Ft Simpson. But there was no thaw from November till a week since. At Ft Liard there were two thaws of a few days each.

Alas I didn't preserve the parents of those pigeon hawks though I had them both. But I compared them carefully with your description and there cannot be a doubt as to the species (I kept your bird book wrapped up in my blankets all the voyage and so could get at it at any moment). It was not altogether my *[fault?]* I didn't preserve them however as the Iriquois had them in their pots boiling (with a crow!) as soon as I was done examining them. But I'll be more careful in future. I have tried and shall try to get very large numbers of all common things even.

Mr. Ross who has within a few days gone into the collecting at a furious rate, has the side of the mess room adorned with some 20 skins of red squirrels, 10 of *Hesperomys* and several whiskey Jacks etc. I've taught a small savage to skin and Mr. Ross keeps him hunting and skinning all the time.

I hear accounts of a few specimens being collected at various posts. Taylor the Postmaster at Ft Norman and Mr. McFarlane of Good Hope both now here, have recieved lessons in bird skinning and promise to teach Indians and get them at work skinning. Mr. MacKenzie at Ft Liard got to be a good collector ere I left and I hope will send out a fine lot of specimens. He skins quite well and I taught some Indians while there *[blotted]*.

Please read my letter to Prof Turner for information respecting my abortive attempts at Philology. Try his Lordship Bishop Tasche of Red River for Indian Languages. Mr. Mactavish of Red Riv is a catholic by the way. Not a very strong one though!

Mactavish says he sent my seine and copper can of alcohol etc. to Churchill Riv and made a dead failure—not a thing collected—some blunder. He will send it in my "way", he writes, next summer.

Begging Mrs. Baird's pardon I'm not in the least crazy. It is true my beloved snaiks are not to be found here but then, but tis no use telling *her* of musk oxen, inconnus, arctic birds and *arvicola xanthognatus*. Tell her I really believe she is so deficient in appreciation of good things that she wouldn't enjoy a dinner of *back fat* and pemmican with tea that *is* tea. McFarlane and I expect fine fun hunting carabou and musk oxen and drinking whale oil winter after next on the Anderson, and will make a trip of a few days down to the Sea coast; if I stay till summer there I'll bring Stim arctic marine invertebrates!

(see other paper)

As for Dr. Lecontes astonishingly liberal offer to pay $25 for the pick of the insects I'll see him "on a long voyage with mal de racquet, short of provisions, and in a *thaw*" before I'll accept it, so far as I'm concerned. I'm fully prepared to make collections in entomology and havn't the slightest idea of their being so insignificant as all that.

The fish from Slave Lake were brought here frozen and too much broken for skinning and now are caught here in winter, but I'll get plenty for skinning in spring. I take my seine with me of course and hope to send out fish by summer boats.

Mactavish promised to send me all the things I asked for so I suppose the *grog* (6 gals of wine, and brandy for Mr. Ross and 6 gals rum for specimens) will come. Mr. Ross is much pleased at this and it has so opened his heart that he says my trip to Slave Lake shall cost me nothing. It would have been a high figure had he charged me for the dogs, man and provisions as he might have done.

He is getting very enthusiastic on the subject of collections and says I may tell you to expect several cases full next summer. He suggests that you should send a gallon of unmedicated alcohol or whiskey *for drinking* to every gallon for preserving specimens and I think myself it would be the best policy !! Nothing would be more acceptable to any of the officers here than the element of conduction, and if we can manage to do the liberal in furnishing them some of this I'm sure they will collect well. Could you not send in a number 6 to 12 small copper cans, say to hold one or two gallons each, for alcoholic specimens these could be distributed to the various posts or to such gentlemen as showed the most interest in collecting. But truly I don't believe money could be better expended in getting specimens here through the officers than by sending in some drinkable liquor. Whiskey perhaps the best for the latter purpose unless you can get them to send Rum from Norway House, or some brandy if it doesn't cost too much at Norway House or Red River.

Murry of Canland Scotland, the naturalist that got the *Someteria v. nigra* from Mr. Ross, has won the hearts of Mr. Ross and the others here by getting from the H.B. Co permission to send liquor "in such quantities as are *actually required* for preserving specimens of Nat. Hist. etc" into Mackenzies River and by telling Mr. Ross to order from York Factory whatever he requires. (He cannot however get it on this account from Red Riv or Norway House and will not get his first supply till summer of 1861) Mr. Ross has ordered 12 gallons, half of which he says he shall drink and use the rest for specimens, considering it "*actually necessary*" to imbibe as much as is used for preservation of specimens.

Under existing circumstances I think I cannot do better than order another supply of rum from Red River telling McTavish to send only about 6 or 8 gallons (for drinking) in case you are sending in plenty of alcohol for specimens. I shall tell the gentlemen here that those only who collect will get the Liquor for drinking. And you must understand that a half gallon of grog is a grand prize here!

Of course the liquor brought in is nominally all for specimens. I see the 5 galls of rum I got at Norway House cost me 12 shillings a gallon, probably it costs more at Red River but I cannot be sure of getting it by order from Norway House as only Mactavish will break the rule for me. But Sir Geo said I might receive liquor for specimens! But stop, I'll not send any order to Mactavish for anything till next fall after I get your letters by the portage boats, then I'll know what you are doing and will have time to send my order by winter packet if need be.

McFarlane says that if I'll be sure to bring a gallon or two of *the element* with me to the Anderson, that he'll warrent that under its influence will kill musk oxen! But thats enough on the liquor question!

As Mr. Ross has taken so much interest in collecting I think it rather better that he should get the larger animals to send you in his own name, of course I'm satisfied so long as the Smithsonian gets the specimens. Some of the things he collects he will send to Murry of Canland, but you will get most of them and really I believe you may expect a "pile".

I have packed and left here ready to go out in spring one case of dry specimens and a two gallon keg of alcoholic spec's. I shall write to Mactavish, when they go, asking him to change the liquor on the alcoholics in case they are to remain all winter at Red Riv. My Ft Liard specimens Mr. McKenzie will pack with his own. None of the larger animals have yet been secured, though Mr. Ross has promises of nearly all the desiderata among these. Taylor, Postmaster at Ft Norman, says he will certainly get musk oxen sooner or later pretty certainly next fall. I offer him a gallon of grog if he will get me a specimen or two in addition to the two Mr. Ross wants, viz one for Sir Geo and one for you.

In sending anything to me by boats leaving Red Riv June 1861 mark such parcels as are to stop at Ft Simpson and such as are to go down to me at Peels River. You might write to Mr. Ross about them and he could open the cases and keep what is intended for him. And as I will know what is in most of them I can give him instructions from the Yukon after recieving your letters up to Oct 1860.

I'll here remark that this packet which leaves Red Riv about Dec 15th gets to Peels Riv in April and sometimes goes down on last snow to Yukon. At other times it doesn't reach the Yukon till June by open water, but always in time to answer letters and send them up to Ft Simpson the same fall. Next spring, I have reason to believe, the letters by winter packet will reach me on the Yukon in May.

My father enclosed me a couple of extracts from newspaper referring to my expedition. They are d___d "canards" and place me in a ridiculous position. I fear despite my directions to the contrary that my letters home have in part been published. Much that I write in a private letter referring to the H.B. Cos affairs had better be left unsaid in public. I hope this won't result in getting me into bad odor with Sir Geo Simpson if he comes across some of this d___d nonsense, but am afraid it will. I perhaps take needless pains in begging that you will let no one publish any of my letters sent you.

I see Mr. Ross is writing to you. I trust you'll give him the full benefit of your persuasive powers!

I finish this in the night of 25th, am all packed up to start for Slave Lake tomorrow morning. Will try to write you from Big Island to say how I get on and whether I go to Ft Rae.

Please give my kindest regards to Mrs. Baird, Lucy and Gen. and Mrs. Churchill and to Prof Henry's family, and good words to all the other friends.

Consider me booked for a third winter if Sir Geo. consents—for two winters, and the second on the Yukon, *anyway*.

Please don't be in dispair till you hear what I do on Slave Lake. Up to this time I've done no *grand chose* but I keep a good heart in the expectation of better results next summer. If I fail then I will. I'll stay till I *do* get a "pile". Yours always

R. KENNICOTT

13. BERNARD ROSS to SPENCER BAIRD

Fort Simpson, McKenzies Riv:, 25 March 1860
[SIA, HBC Corr Coll, Folder 41]

Dear Sir,

I was in the 22d Inst. put in receipt of your note of the 3d June 1859, but as I have already written you upon the subject of the *Somateria v nigra* further than to state that the bird does not breed on Gr. Slave Lake, and from its extreme emaciation was evidently a wanderer, I have therefore little hope of being able to procure you a Specimen.

It is probable that I will be able to send you several cases of objects of Natural History and Ethnological curiosities. The collection will be distinct from Mr. Kennicotts and may be considered a contribution from the Officers of the McKenzies River District to the Smithsonian Institution. I can skin birds, and mammals tolerably well now, and careful and accurate measurements will accompany the remarks—on their habits, migrations, and pairings. Mr. Kennicott proceeds to Slave Lake but I will make a tolerable substitute here though the want of preserving materials will rather militate against my usefulness.

You will see by the notes I sent in winter that I had, prior to Mr. Murrays discovery, a suspicion that 3 kinds of Canada Geese (including the Hutchins) would be found, and I also imagine that 4 species of Snow Geese will be made out including the Blue variety which does not pass here. I have had a good deal of experience with the water fowl from my sporting propensities, tho' my present situation is not a good one for them. I remained 6 years at Gr Slave Lake where I had ample opportunities of observation.

There is no bird here to my knowledge here with the habits of the English sky lark that I know of.

There is every wish to aid in forming your collection. This pursuit is an amusement to people whose stores of entertainment are not overly large. My own mind lies rather to Ethnology and meterology: both of which I have studied a good deal: but I find that I take very kindly to *skinning*.

I am in hopes of sending you out next Season the Musk ox, *perhaps several*. Goats, Sheep, Bears, wolverines, a white and blue Fox—the latter two I have purchased from the Company myself—you can send me a few books for them—Marmots—besides *every thing* that we can see.

The greatest present you can confer on the Gentlemen is to send in a good stock of spirits for preserving *one half not medicated*—as we must get liquor in sub rosa—with this *stimulant* there is no doubt but that you will obtain lots of things.

I am keeping Mr. Kennicotts expenses as low as possible. He pays for nothing except his personal necessies, passages in our boats and with our winter parties are granted him free, and the *[population?]* of the Forts are always at his service.

There is one thing I wish to mention, that I will consider the *second best* duplicates of all specimens sent *by us* as remaining in the keeping of the Smithsonian for myself if I ever should claim them. *The best* is for your Institution and the others also to despose of as you like. The larger mamals Mr. Kennicott would find difficult to procure. With me it is merely a question of time, and I am determined to have them coûte qui coûte.

The greatest favour you can confer on me, if you wish to send me anything, is a few books. I am a *reading* man and have already a well selected library. French and English books are quite of equal interest to me, as I speak the former as fluently as the latter language.

I will conclude this Scrawl by assuring you that every means in my power; not small in here, will be put in requisition to obtain what you require. Write fully to me on all subjects, any questions that I can answer I will do so *correctly*. My opportunities of observation have been large, and altho my other pursuits have not allowed me to pay much attention to Zoology, still I have a slight knowledge of it. I am getting out a potographic apparatus and intend to copy animals etc. from nature with it, and as I am a tolerable sketcher in water colors I will color these after Nature.

In haste, I remain, Dr Sir, Yours most truly

BERNARD R. ROSS, C.T.

14. BERNARD ROSS to SPENCER BAIRD

Fort Simpson 20 June 1860

[SIA, HBC Corr Coll, Folder 36]

[Professor Brandt collected specimens for the British Museum. He was a resident of St. Petersburg (Petrograd), and worked on the mammals of Siberia. The *Canadian Naturalist* was founded in 1856 by Elkanah Billings, and was the first "Canadian" scientific journal. It was published by the Natural History Society of Montreal, which became one of the first scientific societies in Canada when it was established in 1828. James Anderson (1812-67), a native of Essex, was a Chief Factor and in charge of the Mackenzie River District just prior to Bernard Rogan Ross. Illness prompted Anderson's departure from the north in 1857. Although he planned to retire at that time, he took furlough instead, accepting an appointment to Mingan in 1859. In this letter Ross refers to Baird's "works" on mammals and birds and to a pamphlet by Andrew Murray. Murray wrote a series of articles called "Contributions to the Natural History of the Hudson's Bay Company's Territories" published in the *Edinburgh New Philosophical Journal* (1858-59). Murray's article on Reindeer and other mammals appeared in April 1858 and his article on birds appeared the following April. Baird's "birds and mammals" are the two volumes (birds, 9; mammals, 8) in the *Reports of Explorations and Surveys to Ascertain the most Practicable and Economical Route for a Railroad from the Mississippi River to the Pacific Ocean* (Washington, D.C., 1857, 1858) that he compiled. The data in these monographs was subsequently republished in *The Mammals of North America, The Descriptions of Species Based Chiefly on the Collections in the Museum of the Smithsonian Institution* (1859) and *The Birds of North America, The Descriptions of Species Based Chiefly on the Collections in the Museum of the Smithsonian Institution* (1860). Sir William Edmund Logan (1798-1875) was the first director of the Geological Survey of Canada.]

My dear Professor

I have just finished addressing six cases of Specimens to your Institution, a list of which is forwarded with this, but under seperate cover. And I trust that they will reach you in safety.

The collection embraces specimens from Fort Simpson, Lairds, Halkett and Big Island, the latter not yet seen: but which I will pack when I pass the establishment. As yet there are in all 507 articles, 450 of which are from Fort Simpson.

My labours to make this very large collection, *in the space of two months*, have been very great. A want of tools, cotton, and above *all of arsenic* has caused me much toil, as it was only by stoving and frequently changing the stuffing of the animals that they could be preserved. The stoving has rendered the skins so brittle that I would recommend you to be careful how you handled them until they have been damped and a preservative applied. Owing to a mistake of my *deputy skinners*, I found that a considerable number of the smaller birds have not been skinned down to the wrist joint; but as I expect that the stoving has thoroughly dried the flesh, there is not much chance of their spoiling. You had however better examine them at once on receipt.

The total of the measurements have been taken by myself with every attention to accuracy, and the general operations overlooked and about 100 of the Specimens skinned by me. A considerable number were destroyed from want of arsenic, although in some cases I liberally used spirits of turpentine which I found of service though it slightly spoiled the look of the feathers. I had strychnine but no alcohol of Sufficient Strength to dissolve it, and a small quantity of corrosive sublimate which I used on the dog and wolverine. A want of alcohol prevented me from preserving embreyos and fish, though I skinned some of the latter. The small birds are greatly broken on account of the largeness of the shot, but I have done my utmost to have them well done, and with the means that were at my command I think I have succeeded tolerably. I had no cork nor net for Butterflies of which there are a great variety at this place. In the article of Eggs I have not been so successful as I could have wished. As to making the collections myself, my eyesight is neither so good nor is the country adapted for *pleasure*. You can have, My dear Sir, no idea of the horrid thickets of willows with which this Island is covered and in the penetralia of which the birds erect their habitations—then the moschittoes! from those torments Good Lord deliver us! Their name is legion. I was therefore constrained to employ the Services of the Indians and children for the purpose, who brought me certainly a considerable number, but not identified but as I only pay half the price for Such, they will in time become more careful. Poor as the collection is it has cost me a considerable amount to make it: but I wished to make a beginning, and next year will lower the price. There are immense numbers of the small birds who nest here: but the willows are so thick that they are very difficult to discover. Let us however live in hope! I have however procured two or three of your wantages, vz. those of the widgeon, Buffle head, stock duck and a number of Gulls not identified. I have been informed that the wax wing builds only on the tops of the loftiest tress and that its nest had been discovered by accident by my informant on cutting down the tree, and he holds out little hopes of its being obtained. The Eggs of the Canada Jay will be more easily procured though still difficult. I can at least hold out good hopes of obtaining the eggs of all the water-fowl from the Natives, and which they will identify with ease and accuracy.

In the cases are a goose and a gull, which I am strongly disposed to consider new species. They do not now show exactly the same markings as when alive. The fawn color of the goose's belly was much brighter, and the neck of the gull was distinctly washed with yellow. If you consider these birds as undiscribed as yet, I should wish them to be called *Bernicla Barnstonii* and *Larus Andersonii* after two esteemed friends. Among other rarities are a wolverine, two hoary marmots (*Arctomys Pruinosus*) and a large number of *Plectrophanes Picti* in full plumage. From your book I saw that your Institution was in want of perfect Specimens of these birds so I collected rather a large number of them, as they were, for a wonder, exceedingly plentiful here this Season. You will from the list of measurements see the variations on the tails of the Geese, which contain from *14* to 18 feathers, but which are in other respects Similar. Mr. McFarlane writes me from Fort Good Hope that he saw a specimen of the *Bernicla Leucomelia* there this spring, but which was too much broken for preservation. A set of white and blue Foxes are forwarded. The prime skins I was obliged to purchase myself from the company at a cost of 26p. The common I sent on my own responsibility gratis, though I fear that the step if known would not be approved of. I would therefore recommend you to apply to Sir G. Simpson for permission to have a complete set of the common furs sent you in various stages of pelage, which, as they are of little value, he would doubtless grant.

I would feel gratified if you sent me a corrected list of the names of the Specimens sent in these cases, with my own name where different along with the correct one; as by that method I will recognise the animal again. I fear that I must have made many mistakes in the Birds, if not, the fact will speak volumes for the accuracy and clearness of the descriptions in your two works, which you were kind enough to send me, as they have been my Sole Guide. Indeed without these books I am very doubtful if I should have made so large a collection, as the pleasure which I had in endeavouring to identify the Specimens, was of no ordinary stamp. The work on Mammals I have carefully perused on points which I have a pretty intimate acquaintance with, and I was perfectly astonished at the correctness of the whole. In drawing up a paper for the Natural History Society of Montreal on the fur bearing Animals of this District, I availed myself liberally of your aid: but I trust in my quotations you will absolve me from plagiarism when I tell you that in all cases I have compared your discriptions part by part with either freshly killed animals or with skins from the store. The Series has reached only to the Bears but I intend *(DV)* [*Deo Volente* or God willing] to complete it next winter. It grew to such dimensions that I found my time too circumscribed to finish it. The quantity of Skins which I examined in writing it was immense. In the paper on martens I passed above 200 under review, and I rather incline to the supposition of Prof Brandt that there is a cross of the Sable in the N.W. Martens, ie. those from the Youcon and the western Arctic coast. Should these papers be printed in the *Canadian Naturalist*, I hope that you will see them, as I should be *most happy* to receive any advice or instruction from you on any of the points mentioned or hints as to what to particularly direct my attention. My self love you cannot hurt as I am perfectly aware that I am as yet a mere neophyte on the threshold of Natural Science, And am perfectly willing to be introduced into its precincts by so illustrious a notary as yourself.

Accompanying this are a few notes on the Mammals sent, it was my intention to have sent a few remarks on every species forwarded: but the multiplicity of my avocations prevented me. I hope however in time to pass the whole fauna of the District in review in the style of my treatise on the fur Animals, with colored photographs of every species discribed.

Owing to a scarcity of packing cases there yet remain a considerable number of specimens which must stop until next Season, besides about 18 Birds Nests, Some Butterflies and beetles. If I get a good set of instruments and a supply of preservatives etc. I will do my best to work out the Natural History of this place, as well as to influence the officers of the District to do the same at their posts. I have bestowed every attention to packing them carefully: but the road is long and rough beyond conception.

I received from a Mr. Murray of Edinburgh a few copies of a pamphlet published by him on the Natural History of the Hudson's Bay Territory. I have never in so Small a work seen so many errors and many of them gross. In his article on the Raindeer he is perfectly wrong throughout, and is so moreover with plain and positive proof, which he quotes and then ignores because the reality does not coincide with his preconcieved opinions, staring him in the face. The Barren Ground Raindeer most certainly use their feet and not their horns to procure their food, as at the time when these implements would be required, if such were their use, they have already fallen off the males who shed their horns in December before any quantity of snow has fallen to render Mr. Murrays "plough and shovel" necessary. The females shed theirs late in April or in the beginning of May: as they are furnished with horns though the female Moose have none. The Barren Ground species is *decidedly* distinct from the Strongwood. The average weight of one of the former exclusive of skin and offal is about 100 lbs and of the latter about double that. The two kinds do not intermix, nor are the Strongwood ones more gregarious than Moose, whom they much resemble in their habits, and in similarity of food. The hide of the Barren-ground is of lighter color than the other species which also never migrates to the sea coast. As to the difference, if any, between the Lapland and these deer I can say nothing, but I cannot admit that the non-domestication of the latter is any ground whatsoever for drawing a specific difference between them. He says "that any animal which can be domesticated will be found in such a condition, as the animal invariably makes the first advances." I have never yet heard of wild horses seeking the bridle or of wild cattle courting the yoke. He does not take into consideration the idosyneracies of the human races inhabiting these wilds, in the Indian branch of which, at least, a very prevalent superstition would prevent any attempts at taming, which is in my opinion the first step to domestication—it being an act, while the latter is a process. I have myself little doubt of the perfect facility with which the Raindeer could be domesticated, but the process would require many years to finish—as in a similar manner it requires Several generations to civilize Savage man, though I much doubt if Mr. Murray would agree that there is any distinctive difference requiring a *seperate Adam* between even the Negro and the White. He also states that the red-shouldered Blackbird (*Agelaius Phoeniceus*) does not come farther North than the 57 parallel, it has been observed about the 63°.

Some fossil bones were sent last year, which I learn from a letter received from Mr. Anderson my predecessor in the charge of this District were intended for Sir John Richardson, who has been inquiring about them. If he write to you to request them and should wish them to be sent as they were forwarded through mistake: but as much larger and more perfect deposits of these bones have been lately discovered on the Youcon, I shall take care that the Smithsonian Institution is no looser by the transaction.

I have made arrangements at my own expense for procuring full specimens of the various Marmots, of which there are four species in this District—and at the Companys expense to obtain the R.M. Goat, Musk ox, and Barren Ground Bear—and I trust that the hunters will be successful. Write fully to me of all your wants, and if I cannot procure them (*fur animals excepted*) they will indeed be difficult of obtaining.

The collectors of the present season have been myself, and Messrs. McFarlane, McKenzie, Reid and Dunlop. Mr. McFarlanes collections, owing to the want of preservatives were so full of maggots that they could not be sent except a few of the Small birds which were dry, but these specimens have been so badly arranged that they will I fear be of little use to you and I have therefore neither named or catalogued them. If you could forward Messrs. McFarlane and McKenzie each copies of your *Mammals* and *Birds* you would I am quite sure add greatly to their zeal, and some light publications or newspapers would I am certain gratify Messrs. Reid and Dunlop. I would also suggest the propriety of furnishing a good supply of preservatives without which they cannot collect. Mr. Onion has commenced a very complete and carefully arranged botanical collection which from want of room cannot be sent this Season. A good work on Botany would be acceptable to him, and some papers or novels might be also sent to Mr. Flett who keeps the meteorological register at this place. I need not say that any addition to my own Library will be most acceptable, as I am a complete book-worm. Scientific works published by the Smithsonian etc. but this would be, provided they are not sent from your private means, as I should not wish to be a *tax* on any one. Could you inform me of the probable cost of a Second-hand copy of Audubons *Birds* in sound bindings, as paper covered works are of little use here from their fragility. Any works sent therefore please have them bound, which expense I will repay you. A supply of *unmedicated* alcohol, *marked poisoned* will be acceptable and all favours I shall endeavour to reciprocate to the utmost of my power.

Mr. McKenzie removes this Summer to Resolutn. and I have given him a supply of Shot and turpentine to carry on the war at that place.

Mr. Kennicott is now at Fort Resolution where, from what I know of the place he should make good collections of wild fowl, mice and eggs. I have reduced his expenses for the past Season as low as possible having only charged him for articles out of the Store, and he has had facilities fully equal to those which cost Colonel Lefroy £2000. While here I will endeavour to let his mission cost as little as possible.

Enclosed in one of the cases is a parcel of mocassins to your address, which I hope you will honor me by accepting. I fear that the pair of small shoes are not fellows and I am unwilling to open the case I therefore will send the odd shoes which if I am correct in my supposition will give an additional pair for Your little girl.

I wish you to send a complete set of the *duplicate* specimens contained in the 7 cases sent by me to the Natural History Society of Montreal in my name, and you will confer an additional favour if you will name the Specimens correctly previous to departure. Any expense I will settle for with you, by crediting Mr. Kennicott with the same here, and if you can procure me Audubon's Work at not too high a figure I will repay the cost in a like manner.

I have nothing further to say at present: but hoping to hear of the safe arrival of the cases. I remain, My dear Professor, Yours faithfully

BERNARD R. ROSS

P.S. I have taken the liberty of addressing a case of fossils and minerals for Sir W. Logan to your care. Will you be so obliging as to get it forwarded and send me a note of the expense which I will carry to Mr. Kennicotts credit. There are also three small parcels which you will oblige by getting forwarded to their address.

You may go so high as £30 Stg. for the *large* copy of Audubons work, and if you can procure it send it as soon as practicable addressed to the care of the Hudsons Bay Cos agent at St. Pauls, Messrs. Birbank or by any other *safe Channel.* Everything should be sent in soldered tin cases within wooden ones, which will be available for sending specimens to the Smithsonian.

15. ROBERT KENNICOTT to SPENCER BAIRD

Fort Resolution G. Slave L. June 29th 1860

[SIA, RU 7215, Box 13]

[William Lucas Hardisty (c.1822-81) was the Chief Trader in charge of Fort Resolution when Kennicott visited there in 1859-60. Hardisty was a Rupert's Lander, the son of Chief Trader Richard Hardisty and a native woman. William worked for the HBC from

1842 to 1878, spending his entire career in the Mackenzie River District. Mrs. W.L. Hardisty was a woman named Mary Allen. She was probably a Kutchin or Loucheux woman since Hardisty was stationed amongst these tribes during the mid-1850s when they married. (Hardisty to George Simpson, 10 Nov. 1857, H.B.C.A., PAM, D.5/45, fo. 263-264). Lawrence Clarke (1832-90) joined the HBC in 1851. He was born in Fermoy, Ireland, but had lived in the West Indies before going to Rupert's Land. He became a Chief Factor in 1875, and he was active in municipal and territorial politics in the 1880s. Robert Campbell (1808-94) left Scotland for Rupert's Land in order to assist in establishing an experimental farm at Red River. In 1833, three years after arriving in North America, Campbell left the farm for a clerkship in the fur trade. He was sent to the Mackenzie River District in 1834, and spent the next twenty years exploring throughout the north. He was given charge of the Athabasca District in 1855, and remained there until 1863. He married Elleonora C. Stirling in 1859 and they had three children. He spent the last eight years of his career at Swan River. Kennicott refers to two instructional pamphlets in this letter, namely; "Circular in Reference to Collecting Nests and Eggs of North American Birds," Smithsonian Institution *Annual Report* (1858) and "Directions for Collecting, Preserving and Transporting Specimens of Natural History," Smithsonian *Misc. Colls.*, 34, (1859). William Sinclair (1794-1868) was in charge of Norway House and the Norway House District between 1858 and 1862. Sinclair, a native of Rupert's Land, joined the company in 1808 as an apprentice clerk. In 1850, over forty years after joining the service, he became a Chief Factor. He retired to Canada in 1863. Philip Reese Uhler (1835-1913) collected specimens for the Smithsonian, acted as librarian and assistant to Agassiz at the Harvard Museum of Comparative Zoology, and became the librarian of the Peabody Library in Baltimore. Isaac Lea (1792-1886) was a Philadelphia businessman, publisher, and malacologist. He described more than 1800 mollusks during his scientific career, studied geology, and was the President of both the American Association for the Advancement of Science (1860) and the Academy of Natural Sciences of Philadelphia (1858-63).]

My Dear Prof Baird,

I delayed writing to you until I should have done something at the collecting, and since I've been at the eggs I couldn't *tear myself away* even to write to you. I'm sure you'd rather see a nest more and a little shorter letter. (July 5th - I'm *sure* of it now. You'd rather have it shorter anyway!)

The register must tell its own story. I have been forced to copy it in a hurry from my notebook, but have re-compared all the measurements and there are no mistakes in these.

Podiceps auritus, *Grus americanus* and *canadensis*, *Colymbus arcticus* and a n.sp. *Anser albatus* or n.sp. Eggs of *Stercorarius*, *Zonotrichia gambelii,* widgeon, Green wing teal, *Spizella pallida* etc. etc. Plenty of *Arvicola xanthognathus*.

Let the mention of above species serve as a sort of sweet coating to the bitter pill you must now swallow. Though I have a few good things, this springs collection must be put down as rather a failure. Yet I have this rather poor consolation that it is because I have been very unsuccessful in finding them, and not because the things to make a fine collection are not here. As yet my attention has been given almost wholly to birds and eggs, perhaps too much so as I find the boats are almost upon me below. I have once tried for small fish. But it is only a little over a week since the ice was at the shore of the lake and prevented my getting off to any river for fish. You see by the register when I began getting eggs, since then I've done little else than hunt eggs, not even skinning the parents, except rare ones, but having it done by an assistant whom I have taught. I have great disadvantages here that I did not appreciate till I experienced them. The country near the fort is almost impenetrable in spring. All swamp where tis not impassible thickets, and worst of all the Indians are the most unobliging imaginable and as big rascals as such idiots *can* be. I can get scarce anything from them, and what few specimens I do get I must pay large prices for. So long as an Indian isn't hungry, or in fact *very* hungry, he is as independent as you please and scorns the idea of working for anything less than very large pay; if he will work at all. But no other Indians in the District are as bad as these. At the Yukon I'll get them to work well Mr. Hardisty says.

Still I think this is a better place than Ft Simpson tho! I would perhaps have done more at Ft Rae. I came here more especially on account of water Birds relying too much on Mr. Ross' representations, and as you see of water birds eggs I've scarce got any good ones. (Though I shall in a few days go to distant islands for gulls eggs) Of Insects, shells, plants, and fishes I've gotten few because I've scarce spent any time on them. But may do well yet.

I ought to have staid at Ft Rae till the boat came here, which it did 3 days since. But 'tis no good giving reasons *why* I've not done better. I was lazy in the winter it is true. But I can honestly say I've worked as hard as I could since the birds arrived.

You see by the Register what Clarke got at Ft Rae. It is from it I judge that I should have done far better there. I hope to see a good deal from Ft Simpson and Big Island and something perhaps from Good Hope and Norman. Mr. McKenzie unless he has been starving this spring may send me a good lot.

I don't know what you'll think of the collections but for myself I'm bitterly disappointed. Still I'm not going to feel dispirited at it; but just try the harder the next month here, and after at the Yukon. If I fail there (but I *won't*) why I'll get the more musk oxen and ptarmigan's and white geese eggs on the Anderson and perhaps Arctic coast. If I can't get them or still fail if I do, I'll get a big collection at Athabasca the year I go out *or break something*. In short, far from intending or thinking that the present failure will extend to the whole expedition I am more than ever determined to get a big collection "or perish in the attempt" as the heroes of novels and tradigies say. But mind you I write this very good humoredly, smoking a large stone pipe, and don't feel at all tragic, but just used the above as a strong form of expression. Please tell me just what you think of the collection (You may I think generally rely on my identifications of the species) judging from the register if you don't get the specimens this fall. Now please don't, to encourage me, make it out better than it is. By the time I get your letter

I'll have such a collection as will quite reconcile me to it if you say this is even worse than I think.

I must say that a good many of the later bird skins are poor though in good plumage; as I couldn't possibly skin and measure them all myself. But I believe my own skins are not generally bad. *I cannot skin rapidly enough*, a half hour to even a small bird is the best I can do. I have taught an Indian woman (I'll mention in self defence that she is about as good looking as old McPeak) to skin birds pretty decently, tho she makes no really good skins, and have her skin those I haven't time to do, nearly all of the parents of eggs are her skins.

For the reasons I have given I find egg hunting slow work. The musquetoes are so exceedingly numerous and horribly voracious as to prove a serious detriment to collecting. I think I've lost some pounds of blood during the last week by them. But I could endure their bites if they would only keep out of my eyes. I've lost several fine birds and oppertunities to kill parents of several nests, by their flying into my eyes and destroying my aim when about to shoot. If I try to protect my face I cannot see the birds so readily. The fact that I was wet to the waist nearly every day this spring, for tis impossible to avoid wading through deep swamps and lakes, might have been a stopper to my work but for my unusually good health. I think I needn't fear sickness in the north after this.

I havn't time to copy the few notes I've made beyond what I've included in the register. *Grus americanus* is not rare here though I could only get one. (For all these rarities, big swan etc., *I offered very large prices but these D___d Indians always refuse to do just what you most want them to*) *G. Canadenisis* (if it is is = very small) common. The small snow goose is certainly different from *hyperboreus*, very likely *cassins albatus*. Unluckily I could only get one specimen not very badly mutilated, and had to take that almost by force which I would have done you may be sure rather than not get it. I only saw small *Berniclas (hutchinsii?)* at one time and didn't skin as many as I ought because they were bad and the Indians promised to bring in plenty. But this small one is very common on the Yukon Mr. Hardisty says, where by the way there are *none* of the little white goose. Mr. Hardisty says that he will get for me as many of these easily distinguished birds as I like next spring, and Clarke can get still more as his Indians will do just what he tells 'em to. The *Colymbus arcticus* and *Podiceps auritus* are *all right*. One of the loons is certainly distinct from any in the book. But of two I can't make out which is *torquatus*. One is a third smaller than the other tho being slender tis not proportionally shorter. I have seen no small waders for the last two weeks except spotted sandpiper and *gambetta flavipes* of which I've two nests each. Not a bright lookout that on the wader question. I have several nests of small birds still in the woods, may add some to the packages made up and may send a few fishes if the boats don't come too soon.

At Present The dry skins, eggs, nests and insects are packed in 4 boxes, one a cassette, a sort of "Norwest" trunk. (that is, the specimens collected here and at Ft Rae and most of my Ft Liard specimens which came here July 4th) I left a case full of skins and a 2 gal keg of alcoholics packed at Ft Simpson. I have nearly two gallons of alcoholics here ready for packing but expect a suitable keg from Ft Simpson for them.

The dried fish skins are curled up close but will resume their shape if soaked in water, them for fish skinners. If a fish is allowed to lie till the scales dry and fat melts next the skin, the skin can be taken off without loss of scales; and, if fat, very easily.

The Inconni is not like a true salmon a bit. The colors are very bright and silvery even more so than any of the white fish. They are generally considered poor eating. But are said to be much superior to the same in other localities, at Ft Rae and Peels River. At the latter place and lower Mackenzie River they attain great size. (5 feet long and 1 ft deep and better than any white fish or trout Mr. Hardisty says) They are in the Yukon but small.

Several *salmonaids* known in Slave Lake I have as yet failed to get but will try hard to secure all. Some can only be had at particular seasons. Inconnuis spawn this month, latter part, I think. I will observe. Trout are sometimes 4 feet long in Slave Lake. I send dried skins of medium or small fishes only.

I don't think I can afford to go to Ft Rae this summer as it would take far too long, and the boats pass here going down the first days in August. May go a week before the time to Big Island in canoes to get fishes at tail of Lake, hunt the insects and mammals of that half barren country and put Reed (in charge) up to a few more dodges in collecting.

I don't know if I ever explained fully that as I found Mr. Ross very anxious to send all he could in his own name, I agreed that I would teach all I could to Reed and some of the other Postmasters and clerks who were to hand over to him all the specimens to be sent out in his or his and the collectors names. As I of course wanted to see all the specimens sent, possible, I thought this better policy than to have them given to me. The more so as Mr. Ross rather insisted on it and agreed to pay any expense. But I find almost all the gentlemen opposed to this, all I've seen since prefering to give me the specimens directly. Clarke says he'll see him d___d first and me too! As he says Mr. R. "is too fond of getting others to work and he getting the credit." So I'm afraid will not make the thing work. But I'll try to keep Mr. Ross satisfied and hope he will himself collect a good deal. I shall try to get all the gentlemen to send specimens direct to the S.I., that is such as will not send them thro' Mr. Ross. And I think it would be highly advisable under the circumstances for you to write directly to such as send out specimens this spring. Anyway to Wm. Hardisty Esq, Ft Resolution, Slave Lake, Mr. Laurence Clarke, Ft Rae, and Mr. Alex McKenzie, Ft Liard. From all of these three I've recieved the greatest personal kindness and every facility they could grant me. In fact they have each taken far more trouble to help me than Mr. Ross. You see what Clarke has done in birds but says himself he didn't half try for want of proper collecting materials. I'm inclined to think he will yet be as valuable a correspondent for you as Mr. Ross. All three are men of education, and gentlemen. Clarke Irish, McKenzie and Hardisty natives with some Indian blood. Just please send 'em some books and *write to them*. Clarke protests he don't care a fig for the credit of working for science but wants to oblige me—says he'd be glad to get books tho but will continue to collect any way—is to get me a *lot* of *carabou* next fall. J. Reed of Big Island, who will probably collect a good deal, is a postmaster raised from a common man and uneducated, but very conceited, tho obliging and not a bad fellow at all, and a letter from you would stimulate him to greater exertions.

What Mr. McFarlane and Lockhart will get remains to be seen, not much probably till I have worked at their posts and taught some of their Indians to collect. But Both have been very obliging and If I visit them will be doing very much for me. If you can write to these it would also be well. Those besides Mr. Ross who ought to recieve some books are first Clarke, 2nd McKenzie, 3rd Hardisty, 4th Lockhart and 5th McFarlane. *C. Gaudet* at Peels River, a postmaster and uneducated but really a gentleman and an excellent fellow, will be very apt to collect a good deal through his Esquimaux.

The gentlemen say that Mr. Ross will "gobble" all the things himself that are sent in to the officers generally through him. So pray if anything is sent to be distributed, state what is for Mr. Ross himself. And while I'm here better let me have the distribution of the things myself.

As I've said this much about Mr. Ross I might as well add that I find he has been humbugging me most confoundedly. He made me believe that my presence at Ft Simpson was entirely unprovided for and that I was living on his allowence. I now find that he had left at Ft Simpson far more than a full allowence for me, over and above his own, Onions and the Parsons. So I am not *personally* under as much obligation to him as I thought. But please observe that I mention this only as a personal matter. I have nothing to complain of so far as his facilitating the objects of the expedition are concerned, and I am in hopes he will himself collect well for you yet. And after all these little exhibitions of meanness may be more than made up for by his collecting operations. I believe he will always help me too, when it doesn't interfere with his interests in any way. I wouldn't have mentioned this at all if Clarke and Hardisty hadn't urged me to, and as I at first, relying on his statements and promises, said to you so much about my sense of obligation to him, I thought it better to say so when I found I was mistaken about living on his allowence. I shall manage to get on very well with him though now that I'll not be staying at Ft Simpson. I trust you'll "butther him up" a little in your letters as he can do a great deal and I think intends to. I'm all right with all the other officers and as I've said, get the most pressing invitations to stop at their posts.

A more agreeable subject than speaking ill of "the Bourgois"—an Indian just brot me a very good skin of *Colymbus arcticus* and I just now found another nest of *spizella pallida*, but no eggs yet.

I must ask you to write to one more gentleman. That is Robt Campbell Esq., Chief Trader in charge of Athabasca district and his post Ft Chepewayan on Lake Athabasca. I saw him only a moment tho' I had a letter to him from Mr. Barnston to him. Mr. Hardisty tells me he is a very good fellow and would be apt to interest himself in collecting if put up to it. Mr. Hardisty will speak to Mrs. Campbell on the subject in passing to the Portage. Mr. Campbell goes every summer to Norway House for his outfit and I think it would be an excellent plan to send a little arsenic, alcohol and bottles for insects to him at Norway House, and write to him. He is not a man of any great education, but Mr. Hardisty says would be very apt to collect well either himself or through the officers under him. I've not written to him yet because Mr. Ross and some others with whom he is at bitter war gave me a wrong impression of him. Hardisty has been long with him and says he can readily get me into his good graces. Says he is sure Campbell would recieve me very gladly if I spend the spring there the year I go out.

I don't know if I wrote you before that I now intend to go in the winter previous to my leaving, to Athabasca and stop there till Mr. Ross passes on his way to the portage. Fort Chepewayan is said to be even a better place for water birds eggs than Ft Rae. Clarke who has been stationed there says the Indians always bring in eggs by the canoe loads so that they are given out as prey to the servents. I'll try to get Mr. Cambell to get eggs next spring if he has not done so already. I sent him egg circular and S.I. pamphlet, and Mr. Hardisty who visited him last fall told him what I was up to etc. etc.

I have now given up my first idea of wintering at Peels River. If I do not spend the summer of /62 on the Anderson, or some other unforeseen place, I will leave the Yukon with the boat in August, have a few days at Peels River and LaPierres House and then come up the Mackenzie and reach Ft Rae by open water, stay there to get the more eastern and barren ground style of mammals and winter birds, and toward spring go to Athabasca. If I stay till /63 I'll leave the Anderson in time to be at Ft Simpson before the fall meeting of gentlemen breaks up, then very possibly spend fall and early winter at Ft Halkett and go on snow shoes to Athabasca at such time as I fancy, or after a couple or three weeks come down the Liard and still get to Ft Rae by open water. But if I go to the Anderson I shall care less about Ft Rae. You see it is really quite out of the question that I should leave satisfactorily without spending a 4th winter. I believe the Halkett trip would be first rate. I could substitute for it a winter at Dunvegan but *that* would be less desirable because I would arrive too late and leave too early. Besides I have still a "hankering arter" Jasper's House or some other place in the Rocky Mts farther south, which will do for a trip after I get home.

My expenses here this summer and cost of getting from Ft. Simpson here, including also cost of specimens at Ft Rae and Big Island, will not Exceed £20 if it comes to that. I think Clarke has less than five shillings charged to me for all the birds he got of his Indians. I'm to pay the woman that skinned his specimens but only a trifle. But here I've paid the most rascally prices, still I thot it better to do so than not get the specimens. The more so as I *may* have to go out next year. But I shall know how to manage better next time. I have gotten from Mr. Hardisty of the things most in demand among the Yukon Indians, and of these I can leave a large lot. Mr. Hardisty and Clarke have agreed to let me have a lot of things from them. Goods to the value of a "skin" of some articles is charged to us in the shop at a much greater or less price than a skin of another (But mind this is one of the companys secrets and mustn't be told of) so as I'm now up to the dodge I can manage much better. In short the question of funds doesn't trouble me at all. Tho I want to be as economical as possible in order to be able to have a little to send to each proper post to purchase specimens at these.

I did very wrong not to ask for more little things from home. If you send things after you get this I think it would be very good to send a lot, say 5000 common needles and 200 darning needles, the large size of sewing needles best out, some small—about no 1 and 2, mostly a few no 7., cost about $5 to $2.50. (10 needles are worth as much to me as some articles costing 1 shilling stg!) A few pounds of thread (part of it good linen for my own use) a large lot of cheap, colored silk ribbons and five or six dozen large common *colored* cotton handkerchiefs. The latter should be bright colours (if possible) and 2 feet square. They are called "four fourth common cotton handkerchiefs" I think. Several dozens (say 4 or 5 dozen) common pipe heads of strong material. A dozen or two cheap pocket knives. A lot of tinsel hat cords, sham gold and silver "jewelry" and colored feathers if the latter are cheap. A few colored calico shirts and some cheap bright colored calico if these are not too heavy. But 3 cotton handkchfs are worth as much as a shirt to me. I wish I had a lot of these things. I could get the Indians to bring me lots of specimens for them. Even if the things come after I leave I could send them in to such gentlemen as would use them best in getting specimens. The Company's goods for the Indians are generally of a heavier (and more useful) character than the above, but still the Indians are ready to do anything to get them usually. Tobacco I can get here cheaper than it can be sent probably.

I've been thinking that as the Indians are all very eager to get double barrel guns, it would be a jolly good thing if you could send me a couple. Flint locks are best though percussion locks would do. I suppose they might be had for $10 or $15 each and here they would always be worth double that to me, but better than all I could by offering such a grand and generally unattainable prize, induce an Indian to spend his whole year hunting specimens for me, and I thought with one sent to the Anderson or Ft Norman I'd get several good musk oxen, and with the other at Ft Halkett a lot of sheep, *goats,* porcupines, etc. The Indians can very rarely get such a thing as double barrel gun here and when they do tis considered the greatest of favors. The cheapest kind of a French or German gun would do; and all the better if flint lock. It woudn't be a bad idea even to send one percussion lock gun of decent quality as I'm a little afraid the stock, which is cracking, of mine will not stand. If a good one comes I can always sell it for far more than the cost. The gun I now have cost £15! in England and is in good condition except that the stock is cracked. I wouldn't sell it for $50. I exchanged my krieder gun for it. By the way any guns sent should be about 24 to 26 gauge or not much larger bore. But about the matter of the guns you will judge yourself.

In sending in packages to me it would be well to pack in strong light iron bound boxes made for the occasion, well put together and with hinges and lock. They will then serve to send out in again and for me to use going out. You will of course understand that to have good water tight boxes made thus even at a little expense will be economical in the end. I am forced this spring to send out a cassette worth here £1-5 for want of a common case that is sufficiently tight. Here and at Ft Simpson I packed my most precious small things in India rubber bags brought with me. In the cassete are the next best things or such as are most liable to spoil by wetting. I'll write Mr. Mactavish asking him to open and dry the specimens if they remain all winter at Red River.

By the way can't you buy some light India rubber cloth and have some bags made for me. I find that *two* seams sewed with sewing machine make these quite air tight. Send me also a paper of small *tack* nails for fastening labels and a pound of wrought iron slender nails for fastening up boxes.

I think I can make the deer skin parchment obtained here do very well for labels.

In two of the boxes I pack some Indian shoes. Please give some of these to any such persons (contributors to the fund of Expedition) as would like them, some to Prof Henrys folk if they'd like em, keep what you fancy yourself and send the remainder to my father. I directed one package to my father it contains a gun, coat and fine bag given to me last winter, which I wish to give to my brothers. When I leave I shall carry out a lot of shoes for my own wear (never catch me wearing your corn growers again) and will repay to the fund the trifle they'll cost me after I get home. I suppose this would be allowable. For the shoes sent to my family at home I'll charge myself when I know how many they'll get, and shall credit my father with the cost of a few articles I asked them to send me from Chicago. My wearing apparal used up here must be paid for out of the funds. That will be fair I think.

By the way I hope some arrangement has been made about sending me in a full allowence of tea, flour, sugar, etc., especially tea and sugar. I wish I had ordered more sugar from Norway House. A keg or two of sugar and some rasins to pay small savages for eggs etc. with would be worth sums untold *almost.* I have stopped drinking sugar in my tea and don't intend to use any next winter so I'll have a keg at least, if two kegs come for me. Mr. Hardisty says sugar will buy anything of the Indians at Yukon. Tea I've learned to like as only a Nor Wester can. I find that two lbs a month is really a proper allowence and I want to get all I possibly can as every pound over will be of great value to me in buying dogs, etc. from the men. A man offered me a splendid dog last winter for two lbs of tea but refused an offer of £3 for him from another person. But I know the ropes here now and next fall will manage better. Mr. Clarke says he'll let me buy tea 8 lbs from the shop, and I'll try to buy more if plenty doesn't come for me.

I've not told you yet how I got on, on my Slave Lake trip. Had, as I told you, a hard trip to Big Island. From there went to Ft Rae only walking part of the way. If Mr. Ross had chosen I might have ridden all the way for Clarke had sent a first rate train of Dogs for me, as had Reed from Big Island, and I learned from Clarke that far from his letting me go free being a favor, Mr. Ross had no right to charge me anyway as no men were sent on my account. Clarke opened my eyes to the fact of my having been humbugged by Mr. Ross in this and other respects mentioned. I found too that he had given me the meanest kind of a voyaging allowence. Clarke was excessively enraged and would make a row about it if I'd let him. So it seems that Gov Simpson was not forgetting me at all, or meaning that I was to live on the gentlemans allowence at whose post I stopped. I am of course hurt that Mr. Ross should have treated me so meanly, the more so as he was always *recurring* in my presence to the fact, or rather his statement, that his allowence was short. In my innocence supposing he was entitled to this large amount which I knew he must have kept.

You may suppose that after this I shall not think as kindly of some of Mr. Ross' disagreeable doings, but I'll just keep my opinion to myself and play the hypocrite a little. I shall not get into any row with him under any circumstances, and have made Hardisty and Clarke agree to say nothing of his treatment of me respecting allowences etc. I wish I hadn't begun writing about him, but as I did begin I've given you an idea of the thing lest you should think I had gotten into some row with him or would be foolish enough to do so. He doesn't like me more than moderately well but I shall manage to keep him thinking I consider him *grand chose.* The end sanctions the means the catholic priest here says.

But about the voyage, Clarke offered to send man and dogs across with me, but as I could find no ptarmigans there I only staid a few days and then came across with a man and sled that had taken the catholic priest from here to Ft Rae. Clarke sent a sled of dogs at the same time to Big Island, and I rode all the way in this to the large point just opposite Ft Resolution on n. shore of lake. From here we walked out about 10 miles, slept on the ice and the next day finished the "Grand Traverse" making about 40 or 45 miles in about 14 *hours* walking, exclusive of perhaps an hour and half taken for pipes. But we had to face a gale of wind seldom stopped long. I drove dogs and the Indian picked out the track he had made in coming, as the snow was very hard it only drifted a little and he could find the track by the dogs claw marks. I walked every step of the way and considering the gale of wind and a big robe I had to wear to keep warm, I feel inclined to brag of my long traverse.

I arrived here just about two weeks too late for the ptarmigans. But wait, I'll send out a hundred or two! yet. Clarke brought a few but the skins get so dry that the heads and tails break off. Usually there are plenty of ptarmigans at Ft Rae. He (Clarke) got two skins in partial summer plumage but they are very poor.

In packing the specimens I have pulled out all the stuffing from the larger birds. Is that right? I haven't paper enough to wrap them up properly.

My present principal trouble is want of small shot. Mine is nearly all done tho I've economized it with the greatest care, often neglected shooting birds for want of it. I find I can't make beads work. Many of the parents of eggs I've caught with snares set at their nests. I shall try to *make* some small shot, though shot of any kind is scarce.

For the fat large birds I have mixed soap with arsenic. How will it do? Of course I'd put on dear arsenic if I had enough. I've spent some time and taken great pains to dry all my birds thoroughly and hope they will none be hurt.

Mem. Mrs. Hardisty just called me out to look at 4 pelicans that were passing—the only ones I've seen. They don't go beyond Slave Lake that I can learn. I've several times seen a *Chordulis* (very likely the western *henryi* as I saw none this side of English River.

I can't scare up a solitary shrew, may do it yet tho'. Pray send me a lot of single holed common wooden mouse traps with wire springs. Send a good lot for distribution.

Please look if my Red River specimens are not *Hesperonys myoides.* This is domesticated here as thoroughly as the house mouse below.

I'll probably not get time to send you copy of any of my notes this time. Nor can I write to Prof. Henry, Turner and others. In fact I only write to my relations at The Grove and to you—time too precious even if the boats don't arrive at once. I've gotten no complete lists of Indian words but will get some yet this summer perhaps.

Clarke sends some meteorological registers, packed with specimens, not properly entered exactly. He has taken 7 p.m. instead of 9 and his thermometer was on south side of house though in a box. But 'twas only exposed to the sun at 2 o.c. during a month or so after May 14th when I was there, he moved the thermometer to a good position and entered at proper hours. These facts are noted on each sheet. The Register is from October to June. He says he will do the thing up right next winter. He made the mistakes by entering into a book instead of the blanks.

I've got a pair of young fishers, want to study their habits if I can only get time to observe them a little. They are queer looking customers, about the size of martins now, as gentle and even more playful than kittens. Got deuced sharp claws though. When they are quarrelling they growl, or grunt rather, just like very small puppy dogs. But when pleased chuckle out a note just like that of a migratory squirrel only lower. When hungry they cry out in a curious shrill voice more like the "*zree zree*" of a mink than anything I recollect. But louder and harsher. They eat anything, and are constantly climbing about. They walk very clumsily and don't seem as much at home on the ground as when climbing. I think they must be very decidedly arboreal. They were found in a hole in a rotten tree.

I find the little red squirrel makes big round nests in trees, in which the young are produced. I find plenty of them and have seen the young in 'em.

July 2nd

I've written in this from day to day as I got time or thought of something. I've tried the fishing in vain. Must go to some small river. I find *Hylatriseniata* is really quite numerous here. Am not too sure the *Ranas* are all our species.

I'm writing letters only to the Governor, Mactavish, Barnston, Sinclair of Norway House and to The Folk at Home.

Supposing it should not be allowed me to take specimens of any fur animals could you not make arrangements to pay the full price for Blue fox, barren ground bear, etc. These are *cheap* furs.

Pray send to the Governor some few of my birds and mammals that can be spared if he wishes it. I promised to get him a few. But I shall get more that would please him hereafter probably. Mr. Ross is to get him all his special desiderata among larger mammals and doesn't wish me to send any of 'em, so of course I'm agreeable to that.

Who *is* Murry of Canland. I hope Mr. Ross won't send him many of the new species if there are any. I'll try to get out a list of what is sent him if I can, and manage to secure any n. sps. so far as I know. Tho' Mr Ross promised me to send him nothing but that was unique to his collection. I am in hopes Mr. Ross will send some skins of blue fox etc. I've gotten Mr. Hardisty to leave out of the packs a *blue fox* from east end of Slave Lake, and perhaps Mr. Ross will send it from portage La Loche if the Governor writes favorably about Fur animals.

Please give my kind regards to all of the Washington friends, Prof Henrys family and the *real* megatheria in particular. I cannot write to any of them this time. Tell Uhler, Ulke, Baron Ostensacken and shell men to wait a bit. I'm going into the invertebrates "with a rush" presently. I've never yet had a fair chance at the fish, but shall soon make an expedition to some small rivers. I expect I'll catch it from some of the contributors whose departments havn't been worked up. But one can't do everything in four weeks, which is about the time I've really been able to get at collecting fully, except among birds.

Give my very kindest regards to Mrs. Baird, Lucy and Mrs. Churchill. You can't imagine how much, I wish you would all visit The Grove once. I don't believe it's only because 'tis Home that I think it such a pleasant place. I want my sisters to know yourself and Mrs. Baird very much, and Lucy would like Cora I know. Alice is of a very different style. Both are better girls than ordinary, if they *are* my sisters, but Cora *was* a little *spirituelle* (is that the word) like Lucy.

What has got your brother Tom? Please do make Stimpson or some of the other megatheria over look my silence and tell me all about these old faces that I remember so pleasantly in connection with your little room and the bird room. That is supposing you havn't got time to do it yourself. But I know *you are always* at work and it isn't fair to ask you to write long letters on those subjects. Well when contemplating my long banishment I only sigh when I think of two places—The Grove and Washington—and of the latter only three buildings, the S.I., your home and 299 G. St. Mrs. Baird would add a fourth, but no, " *she* has gone and left me" and I don't care much.

Does Mrs. Baird think me *very* crazy in wanting to go to the Anderson? She thought me so, to want to stay even a second winter here. But think what a good sleep one could have in a night of two months or so, and then the carabou "back fat" and marrow bones! But really I assure you the thing is not so hard when one once makes up his mind to it. I don't hesitate to try the fourth winter. Once I get home I wouldn't like much to come back though.

In sending letters and packages to me hereafter just address them to "Mackenzies River District" Hud Bay Ter" and when they get here they will be sent to the right place. But if you send in things, after you get this, to come by next summers boats please remember that I shall not be at Ft Simpson till after the boats all leave, if at all, in the fall of 1861. If I am to go to the Anderson I will probably meet the Anderson Boat at Good Hope remaining at Yukon as late as I can. Or perhaps I will even stay at La Pierres House till winter, and walk to the Anderson on the first snow having my things sent there from Ft Simpson by boat. But if I go out in 1862 I will *possibly* go to Ft Simpson in time to take the boat for Halket, then make Ft Rae by last open water. However Ft Rae will do later if need be. But if I stay a fourth winter I have no intention of going to Ft Simpson at all the fall I leave as it would only be wasting time. So please write to Mr. Ross what is to be kept at Ft Simpson, and as I will know by the winter packet, at the Yukon, what you will be sending I can send up my instructions.

Of course I cannot yet certainly decide my future course but to recapitulate it will pretty certainly be supposing I have 4 winters—Yukon winter of 1860 and 61 and ensuing summer till August when the boat leaves. Then to the Anderson some time in fall or early winter spend summer of 1862 there leaving in time to be at Ft Simpson when the officers meet, probably spend fall at Halket and go in winter to Ft Chepewayan, Athabasca. I think I'd best go to Ft Rae, however, fall instead of Halket in case I *don't* go to the Anderson. Yet you may give me some advice on the subject. Snow shoe voyaging after a little more experience will be very easy to me.

Let me know your opinion of a summer trip to the sea coast, mouth of Anderson, for birds eggs. If the Huskies (as Esquimaux are always called here) prove peacable I may accompany a party of them to their summer homes on the coast in case I don't find good egging ground on the Anderson. Probably however twill not be very safe and instead I'll hire two or three to devote their summer to collecting and meet me at the Fort before I leave. But don't "let on" to any of my folk at home that I think of this. The Huskies have a hard reputation but if they trade with the company they will take precious good care to keep out of any scrape and I would be treated well. Of course I'd take nothing to tempt their cupidity and promise the chief of the band a good present on our return to the fort. I would get a grand haul if I did go, and I believe tis only a trip of a few days to the coast from the proposed site of the fort on Anderson River.

I've written you a horribly *mixed* sort of letter. But I remember a suggestion you once gave me and have made a cross (x) opposite some of the important passages!

I have just learned that there *are* bivalve shells to be had here, tho I have as yet seen none excepting a single valve of a univ brought me this morning. I may yet get some in time for sending out now; if not tell Mr. Lea I'll look to it yet.

July 4th

The first brigade of Boats from Ft Simpson has arrived but Mr. Ross with the last has not yet come though hourly expected. Mr. McKenzie much to our surprise came to spend the summer here in place of Onion who goes to the portage, and I'm "jolly glad" of it, and very likely Mr. Hardisty will stay too. See if I don't put 'em up to some dodges that will be of use hereafter. Mr. McKenzie got only a few specimens. He might have gotten more but ran into the same error that most do here, that is thought a poor specimen worse than none and *lacerated my feelings* by telling me how he'd thrown away some fine things because he had in some way injured them. However he has got some good things, two porcupines, some marmots, etc. and a lot of rabbits etc. etc. He got no eggs though. Still I think he will do much better yet. He skinned all of the specimens himself—good indication. The loss of his father, mother and one of his children affected him very much and he seems to have neglected anything but the most necessary business. The Ft Liard Indians had not yet brought in any of the specimens they promised me when he left. Mr. Dunlop came down to Ft Liard and Mr. McKenzie said promised to collect. I'll see him next fall at Ft Simpson. He (Dunlop) sent me in McKenzies collection two large marmots and two of the big mountain grouse (*Deh Choh*) I heard of at Ft. Liard.

Mr. McFarlane sent up to Ft Simpson a large collection of water birds but Mr. McKenzie says they were all spoiled and probably one will be sent out, but I'll try to save some of 'em yet if they are not thrown away and any are unique. No eggs I believe though. But McFarlane may get some on the Anderson where he has gone to start his new Fort. So as his failure was not for want of *trying* I think we may hope for something yet as I'll post him *[page torn]* the cause of his birds spoiling, next fall. Reed at Big Island has got a big lot and of good specimens doubtless, as I think I gave him all the information necessary and taught some of his Indians to skin. One woman made very decent skins too.

But now for the best. Mr. Ross has, so far as I can learn, quite thrown my own collection into the shade by what he has gotten at Ft Simpson and best of all he has lots of eggs. As you'll recieve his registers they'll tell the story. The bulk is very great and there cannot fail to be a most splendid lot. By June if this is so Mr. Ross may treat me as meanly as he likes and welcome on condition that he makes such a collection every year. I should feel tempted to tear this letter up and write it all over if I could, if I thought what I had said about him before would make any difference in your sending him books etc. I shall feel quite an affection for him when he arrives! By the way though don't you ever, please, in writing to him about me hint at the possibility of my presence in the district having facilitated his operations any. And indeed I haven't been in any way deserving of any of the credit if he gets a big collection. He wants to send the specimens to you himself without any connection with my expedition, and I'm very glad I encouraged this idea of his though I never thought he would do so well as it seems he has. Of course if he has done so well I will try the harder to get all the others I can to send their specimens through him. But please credit Clarke's and McKenzie's specimens to themselves only *not to me at all.*

If Mr. Ross has made such a brilliant thing of it I will of course see that he gets all the preservitives I can possibly spare. Small shot will be the grand desideratum after preservitives.

Perhaps I'll be able to write more after Mr. Ross arrives though he will doubtless tell you all I could. If he wishes it I'll see that Lockhart makes him a good collection too at the Yukon seperate from what I get. I think Lockhart will do it for him as I believe he isn't one of those who are down on the "Bourgois." But you are the one that must keep Mr. Ross' ambition stirred up. I think if you can only send in the preservitives we'll sweep the north in four years at this rate. Can't you get Sir George to permit the specimens to go *out* to Red River free of freight charges? I fear from what I'm told that Mr. Ross hasn't it in his power to send em free without leave, but perhaps he can still and I think McTavish won't make us pay unless really necessary.

I enclose a copy of what I wrote the Governor this time. The letter to him not quite so hardlooking tho! I thought he might take it as uncivil if I didn't write, though in truth I had nothing to say to him. For the Youkon or die! Yours always,

R. KENNICOTT

I forgot to say that I hope you'll send in strong scissors scalpels etc. for bird skinning and rules for measuring specimens. Rules divided into 100ths of inch are much wanted as these we find up here are bad and very few of em. I've lately lost mine and must now use a carpenters rule. A thin ivory rule, or wooden one, out of a drawing case could be sent me by mail. I'll ask McTavish to try to send me one.

16. BERNARD ROSS to SPENCER BAIRD

Great Slave Lake En route to Portage La Loche, 2d July 1860
[SIA, HBC Corr Coll, Folder 36]

My dear Professor,

I have just closed at Big Island two additional cases to your Institution's address. They contain Some white partridges of Mr. Kennicotts collection, a number of Specimens collected by Mr. John Reid, and a further collection made by myself between Fort Simpson and Big Island. Among other interesting novelties are a good number of eggs some of which Mr. Reid procured: but the greater portion of wh. I found "en route," and among them are eggs of the *Dafila acuta,* or Pintail, *Anas Boschas,* Shoviller Duck (a specimen in box), Brown Crane (1 Egg) and some Eggs belonging I fancy to the *Turdus Palasii,* but as the Parent Bird accompanys the nest you will easily identify it. My contributions in the case are in Eggs—the white sea-swallow or fork tailed Gull (no specimen but it is the same as the blk headed one sent except in color) of this only 3 eggs. The black headed sea Swallow (2 Specimens Sent by me in case) a number of Eggs, about 30—*Falco Sparverius* (5 Eggs), *Dafila Acuta* 8 Eggs (one nest). Mr. Reid has sent besides Several nests of Small Birds, 1 Egg of the Brown Crane, 3 Eggs of the large *Glaucus Gull* (Specimen in case), 1 Egg of the middle sized Gull (*Rissa septentrionalis?*), 1 nest of Eggs, of *Anas Boschas,* 1 Do. of the Shoveller (specimens in case) and of the specimens—1 nest of *Nettion Carolinensis* (specn in case), 2 blk headed Sea Swallows are mine, 1 *Pelionetta perspiicllata* Do., 1 red throated diver and 1 *Colymbus Torquatus* do. I left several gulls which I had shot to be dried and they will be sent next Season. I have crowded our boats a good deal to ship such a number of cases but as this will be the heaviest year I have done it to form a good neuclus for "the McKenzies River Officers Collection." You will therefore be kind enough to tell me what birds or animals you have received specimens enough of, and what you particulary require as yet. At Fort Simpson a large number of birds, mammals and Eggs remained from want of packing cases. They are principally *Lepi Americani, Turdi Swanstonii, Colymbus arcticus* (blk throated Diver), *Setophaga Ruticilla, Dendroica AEstiva, Cotyles riparia, Xema Sabini, Coylaptes Auratus, Lauri Glausecentes, Rissa Septentrionales, Picus Pubescens, Arctomys monax, Colymbus Torquatus, Tringoides macularii, Mareca Americana,* and Eggs of *Setophaga Ruticilla, Turdus Swansonii, Mareca Americana, Bucephala americana, Cotyle Riparia, Larus Andersonii*—in all about *200* Eggs—the names of Some of which I have forgotten. Besides these there are numbers of Embreyos in alcohol, Insects, flies, minerals, fish, in all about a case full.

Mr. Reids collections I had not time to name as I only remained a few hours at the Big Island, during part of wh. time I was employed in packing them. Take care in lifting out the Birds as some of the larger Eggs are among them.

I find out that the wax wing is a *winterer* in this District. I heard after we had passed the spot that the boat in rear of mine found a kingfisher (belted) nest but had not time to dig it out as I was ahead. It was situated in a deep hole in the side of a clayey bank.

I tried to have time to send you a good set of Notes on *the birds* next winter. I have plenty to say about them particularly the waterfowl.

I fear I have quite bored you with my long letters so will *now* for the present conclude by remaining, Yours most Sincerely,

BERNARD R. ROSS

P.S. have you ever enjoyed writing in a tent with about 70 millions of moschittoes biting you?

P.S. Preservatives!
Alcohol (poisoned)!!
Alcohol (not poisoned)!!!
Books!!!!

17. GEORGE BARNSTON to SPENCER BAIRD

Michipicton 26th July 1860

[SIA, HBC Corr Coll, Folder 2]

[In this newsy letter, Barnston refers to several acquaintances, both from the scientific world and from the fur trade. Jared Potter Kirtland (1793-1877) was the founder of Cleveland Medical College (Western Reserve). He was an amateur malacologist and one of Kennicott's tutors. His personal interest in northern North America influenced

Kennicott's decision to go north during the 1850s. John Edward Gray (1800-75) was the Keeper of the British Museum from 1840 to 1875. E. Doubleday worked on the insect specimens arriving at the British Museum between 1842 and 1849. Adam White worked on both insects and crustaceans arriving at the British Museum between 1836 and 1863, and he compiled the catalogue on crustacea (1849). Francis Walker was another of Gray's assistants at the British Museum, and he was in charge of entomological identification and classification. The quality of his work in his catalogues was criticized severely by the scientific community, however. Captain Thomas Wright Blakiston (1832-91) was a British army officer who acted as magnetical observer with the Palliser Expedition. He also sent data and specimens from the Saskatchewan to the Smithsonian in 1858, and he collected zoological specimens in Japan while he resided there during the 1870s. Dr. George Newport (1803-54) earned his living as a physician, but made a reputation as a zoological anatomist. His contributions to entomology were most noteworthy; he had several articles published by the Linnaean Society, and he compiled one of the entomological catalogues issued by the British Museum in the 1850s. He spent most of his life in Chichester and London. In this letter Barnston makes an oblique reference to Louis Agassiz's *Contributions to the Natural History of the United States of America*, (1857-62). Governor Isaac Ingalls Stevens (1818-62) of Washington Territory commanded the northernmost survey undertaken by the United States Topographical Corps to find the best route for the transcontinental Pacific Railroad, and his *Report* was a narrative account of his expedition. The forthcoming maps referred to by Barnston were those drawn by the leaders of the various surveying expeditions, and also included in the *Reports*.

Barnston's reference to Ashley, Smith and Sublet *[sic]* requires some explanation. In 1822, Jedediah Smith and William Sublette responded to an advertisement that had been placed in St. Louis by William Ashley of the Rocky Mountain Fur Company. Ashley was looking for explorers. Smith and Sublette were just two of the men who signed up for Ashley's explorations in the far west. Both, however, played important roles in the field, and in providing arguments in favour of American expansion westward. In 1826, Ashley made Smith his field leader, and Smith explored throughout the southwest and up into the Willamette and Columbia River valleys. While travelling through Oregon all but Smith and two of his men perished on the banks of the Umpqua River. Smith and these two men found their way to the HBC's post, Fort Vancouver. They then went on to Fort Colvile. It was undoubtedly during their stay at the HBC posts that Barnston met the Rocky Mountain Fur Company men. Smith and his partners—David Jackson and William Sublette—later wrote the U.S. Secretary of War extolling the virtues of the regions they had visited, and suggested that the Willamette and Columbia valleys should be settled by Americans rather than by the British. Scientific data recorded by Smith during these expeditions was later used by ethnographers and philologists such as Albert Gallatin and George Gibbs. Peter Skene Ogden (1790-1854), one-time employee of both the American Fur Company and the Montreal-based North West Company, joined the reorganized HBC in 1821. He made six expeditions into the Snake River Country between 1824 and 1830 to "trap out" the area lying between the southernmost HBC territories in the Pacific region, and the United States. Ogden had a reputation for ruthlessness when conducting business, and he was given charge of the Snake expeditions because of that reputation. He was made a Chief Factor in 1834 and given responsibility for the New Caledonia District in 1835. He remained there until his death in 1854. John McLeod (1788-1849), originally from Stornoway in the Hebrides, joined the HBC in 1811. He was a Chief Trader in the Columbia District between 1822 and 1826. He retired from the company in 1848. Janet was one of the eleven children born to George Barnston and Ellen Matthews. Like Janet, Matthews was herself a daughter of the fur trade. Her father was an employee of the American Fur Company, and her mother was a native woman.]

My Dear Sir,

Your welcome favor of the 26th Novr reached me two days ago, and as I have much to write before sending back the Indian couriers to the Sault, I must confine myself to moderate bounds, and seize a spare hour here after for digression. I think I wrote You of the arrival of the Keg of Alcohol. Will it be necessary to medicate it with Tartar Emetic, and will it bear watering, without risk of subjects spoiling in it? The Alcohol I have been accustomed to was generally too strong for every thing, except dark colored *Coleoptera.* I would have had none this Summer, had not Mr. Kennicott favored me with a quart bottle of his own, which I can appreciate, when I consider the need he will have of his own stock removed so far and so long as he will be from all distilleries and Laboratories. I advised him however to provide himself with a Keg of Rum at Norway House, which I do not believe he will forget as he is a true Smithsonian, proved by his liberal Donations to me. My Anxiety to have the Cleveland Supply made me blunder on a Sot's reason for its not coming forward, but all is right now, and I have written to Mr. Simpson at the Sault what was the cause of his letters to Dr. Kirtland not having been answered. I have to thank the Doctor now for the trouble he has taken, and hope he may be long spared in the best of health, to benefit science, as well as to prescribe for the sick.

I have got a few insect pins, but if you send me cork, I will be glad to have some pins along with it, that I may choose the best. I suppose the best way is to have Boxes made here, in which to glue the Cork splints as bottoms. Last fall I ordered a small cabinet with a lot of drawers and small partitions to be made for me, and I expect it in spring, but that I intend to keep for my own family's amusement. When at Martin's Falls I first took a liking to Entomology, as it appeared to be a good Locality for flies. I bred Lots of *Lepidoptera* and a few *Cimbicidae* and *Tenthredinae.* What surprised me a good deal was, my not getting above one male, for 60 females of one species of the Musqueto, that I kept in the larva state until development. If this be always the case it is a curious fact, and may be deemed a merciful provision in nature, in Keeping under what might become *perfect Torment* to the animal world. Send me some of the Pyrelhrune Caucasicum by all means. My Eyes were hurt at M. Falls by a too careless use of the Microscope, which caused me to give up studying Insects; I gave my whole collections to the British Museum, in which perhaps You may consider me to have been too patriotic, but I do not regret that. I found three kind friends in that Institution Messrs. Gray, Doubleday and Adam White, and the same feelings that prompted me then still draw me towards every labourer in Science and every Lover of Nature.

Altho' born and bred in Edinburgh I have already written to Mr. McKenzie to collect *many* things for me as well as that city, some of which are intended for the Montl Nat. Hist. Museum, and I am happy that I now know Your principal wants, as it will enable me to cull for You. Your reasoning on the *basis of Types* of Anl Genera is good, and as far as N. America is concerned, attention to it will produce a better classification. Yet I am disposed to believe that, even perhaps in the next Generation, there will be a *General Convention* of men of Science, a deputation from each civilized Country, to determine not only the best principle on which to classify, but even to attempt the allotment to each species its proper place. I have no hopes however of that occurring in my time. More has yet to be collected and known. I shall send you a bottle or two of the vaccinium the first favorable opportunity per open water, and I shall be most happy to have a taste of the Carolinan Scuppernong. I ought to observe that a portion of Sugar and a few Raisins are worked into our home made, and that a chemist would therefore not find the pure fermented Juice of the Bilberry. I mix it sometimes with port. It then makes what might be called a sweet claret, a very fruity wine.

I have a letter from Captn Blakiston. He was to have sailed in the Sinioan Troop Ship the next day. I am happy he has obtained a stop, but would, like You, have been better pleased to have known, that his destination had been Vancouvers' Island. Hong Kong is a most unhealthy station, and I doubt he may be *fixed* there. He entertains hopes of revisiting America before long. A treatise commenced by him on the "Birds of the Interior of Brit. N. America" will be completed by another Gentleman, and appear in the proceedings of the Zoological Society, but of this You probably are already aware. When next I write to Moose I shall inform JMack of the receipt of his observations by Prof. Henry, and of the desire to have the series in detail, as they were taken at the different hours. At the time I got these observations I thought 26.1 too low a mean for the Temperature of Moose Atmosphere, and I have since come upon a Years' observations of the Thermometer at Albany Factory, and find that I have a mean temperature at that place, (which is to the northward) of 28° of Fahrenheit. Whereas it Should be lower than at Moose My times of observation however were not exactly the same, but 8 A.M. 2 p.m. and 8 p.m. Mr. McKenzie's hours are perhaps better, altho' only very careful hourly observations could prove that, one set at least, in each month. Can Mr. Henry inform me of the three hours which are likely to give the nearest to the true mean temperature of this place, or of any point of observation in the states nearest approaching to it in Latitude. The best hours for observing to obtain mean temperature probably differ much according to the Latitudes, but I would be inclined to think that sunrise for the first observation, 2 p.m. for the second, and 10 p.m. for the third would give a lower mean than that *true* that is, *one* deduced from 24 hourly observations for each of the 365 days of the Year. At 2 p.m. one gets *nearly the greatest rise* of the mercury for the day and at sunrise the *lowest ebb,* but 10 p.m. in the higher latitudes in winter, I should think is too late in the evening, and the scale has shewed by that time perhaps 3/4ths of the descent of the mercury between the two extremes.

Martins' Falls has likely a slightly higher mean than Albany, although only on the same latitude, as near it. By a series of Observations, I found it on that line of our Continent, where the Summer Heat merely thaws through the frozen crust of the Earth, but no more. This is where clay soil exists. In sandy or porous soil and swamp, the surface Frost is thawed through faster and earlier especially where there is no shade from the coniferous Trees.

I am getting so old, and withal so blind that I will make promises to no one, lest I should not have it in my power to accomplish them. I also have seen too many Instruments broken in this country, especially in the carriage, that I cannot ask any from the Smithsonian. If You send anything of the kind, *do not form any expectations.* Although I have *[Laboratory Instruments?]* I will take as much care as possible of them, and work with them to the best of my ability *when Business and duty will permit.* In coming from Fort William to this place, I unfortunately have broken, by malpackage, a Smithsonian Thermometer given to me by Mr. Kennicott. Since Jany 1st I have been getting a Young Gentleman my assistant here, to take observations (three every day) of an old spirits of wine Thermometer, which has lost its colouring so much, that I cannot now with my present Eyesight, read what it marks. It is however a good Instrument for Younger Eyes. I may perhaps give You something next Year on our very changeable climate at Michipicoten if God vouchsafe me health and Strength to get along handsomely. When any Instruments are sent to the Indian country small sets of Instructions should accompany them, shewing how they should be used and how *[illeg.]* are to be rectified etc. Many people can understand a matter by explanation, who would never discover it of themselves.

The *Catalogue of American Diptera* came with the other Books, and I thank You Kindly for it. All that I gave to the Brit. Mus. were handed over to Walker who was so fond of *Chalcididae.* He was glad of them, and I believe has described them very carefully. Mr. Gray sent me the Catalogue Books, all except one, which I *[never?]* could get hold of it. It contains the *sysphidae* being part 3rd of the *Catalogue.* I received two copies of part 1 so that probably some one whom I will never see, for an exchange, has got 2 copies of Part 3. I should like to know if Dr. Newport's discovery of persistent *Beanchiae* in the *Pteronarcys regalis* have been subsequently verified. He was quite delighted with it himself, but I confess at the time I could scarcely think that these superfluctus would be found in every specimen or Individual, it being so contrary to the *General* rule of Nature.

Your last letter gave me so much encouragement that I have unwittingly forgot my first intent of being reasonably prolix. I have only time to say that I thank You again and again for your naming the contents of the small Box, would that it had been 4 times as large with much better things in it! I have no proper vials in which to place small fish, frogs, mice, shrews, etc. send me some half a dozen well stoppled with some Arsenical Soap or preserving liquid, whatever You approve of most for skins of birds and Mammals. I have a beautiful *A. Chrysaetos* for my Lobby, which I was obliged to fill nearly with pepper and salt and camphor before stuffing. The Measure from tip to tip of extended wings was 7 feet. The *G. Dilophus* I ought to have known from the 12 feathers of the tail, but seeing no apparent crest I was disposed to thank it a *strange [illeg.]*. The *Dilophus* must be the common Cormorant on Lake Winipeg, especially near the entrance of the Red River into the Lake. I have seen Cormorants also at the outlets of the Lake, near my late station Norway House. It probably hatches on the flat rocky Islands in the unfrequented portions of that Lake, as It appears always a very shy bird. The Indians here are very stupid; they know nothing about the Bird. Mr. Kennicott I think had one brought to him at Fort William, when I was there. I shall write to Lake Nipigon near the height of land. If it summers there, I think I will be able to get Eggs for You. As for shrews, mice, moles, etc. I have not much hope. I cannot catch them, being nearly mole blind myself. But I shall set the Indian Boys to work, especially if You send a few traps for such vermin. What were Agassiz's contrivances for dredging for Efts, Frogs, Eels, etc? His party were first rate collectors, I am told, (when he visited Lake Superior) I find that I cannot at present purchase Agassiz's *Contributions* neither the Barometer. My packet has bled too freely last Year. What with paying off debts in Montreal, not my own, outfitting two Sons for Vancouver's Island, paying off Instruments on Lots in Victoria, a speculation by which I fear John will never gain anything, and various other Items of extra Expenditure, I am nearly at the stretch of a two Year's tether, and in the sailors phrase, "must haul my wind." Be so kind however as let me have Governor Steven's *Report.* I am often at a loss for a Map of the States, especially those West of the Mississippi, as far as the Pacific Board, shewing the various lines of route into California, Oregon, and Washington Ty. I leave it to Your Judgement to decide whether to send me a mediocre map already out, or to wait for the publication of the *Grand* Map, to be edited from the data given by the exploring expeditions. The first rough Draughts of the country lying between the Snake or Lewis and Clarke's River and the pacific, were made out by me from the computed Courses and distances travelled by my old friends Ogden and McLeod, when they had charge of the Companys trapping parties, they traversed all that territory from the Columbia, south to Pit River and the Sacramento on the West Side, and along the affluents of the South branch, to Salt Lake on the East. The distances and courses were guessed at greatly by these Gentleman, but were down regularly at the conclusion of each day, in their Journals, and I used to compare these with the rough chalkings of Trappers, making as correct a whole as I could of the fragments. Arrowsmith got all these Draughts from the Hudson's Bay Company, and I am anxious to compare his map with the *ascertained true positions.* I never could discover whether what was styled unknown River, after I called Ogden's, from his having been the first upon it, was a tributary of Pit River or not, or whether its general course was from East to West, or from North to South. Smith, partner of Sublet and an Elève of General Ashley's, one of the most amiable men I ever knew—gave me a sketch of the Great Californian Desert, which he had crossed, and also of his route from the Upper Sacramento by *Smith's* River to the Pacific. He took that route northwards, as he failed to discover the passes of the Interior, into the Shasty and Upper Umptqua. From this you will conclude I must be anxious to see a correct map of that country.

I am anxious to see some one take up the subject of the Geographical Distribution of plants in North America. It would be an excuse for me leaving off. I have ever felt that I was not Botanist enough to do the thing correctly or in a sufficiently complete manner. Fur Traders have to work up amidst a host of difficulties, nothing whatever is at hand for them. No Libraries, no Museums, or collections to refer to, no friend near to consult, so that every step he takes is like an uncertain one, and he finds himself at last so full of mistakes, that he doubts whether he can ever be exact or correct, or sure of any thing, until he actually see the same, upon some other authority.

Janet thanks You exceedingly for your gallant offer to assist in shopping, but her Lady friends in Montreal supply her wants so much to her satisfaction that she cannot slight them by going to a greater distance elsewhere. As for music she will not say nay, as she Knows not how long her present stocks and some she is getting from Scotland may please her. She will not fail to make known her wishes, when she has any in this respect, and avail herself of Your readiness to oblige. Navigations will not be open for three months yet, in this Lake. Before then I may have some more to say, and *something more to ask.* In the meantime I must wish You all health and the enjoyment of all true happiness

Ever Yours Sincerely,

GEO BARNSTON

The Marine Algae are most beautiful. I must come some day to Washington to thank You for them which will give me the pleasure of seeing the Smithsonian collections besides. You will soon have a *Tetras Richardsonni* from friend Kennicott, although you may not decide the name as I have done. It may be another. I saw a dark colored grouse like it in the Rocky Mountains Lat. 59" and 53°, but I did not preserve it. I thought it was *Franklinii* (also of Douglas—what made me call it the *Richardsonii* was seeing the white paletus on the neck of the male and the obscure black band near the end of the Tail, and the *smalness* of the female compared to that of the Male. Tis the largest species I have seen of the Pine wood Grouse. Is it the *obscurus?*

P.S. If Bernard Ross's Box do not come forward, Wm Mactavish our

C.F. and Man of Business at Fort Garry R.R.S. will be glad to give You information about it, and to look after it, if it have got astray in the north of the latter I do not myself see a possibility. If it be missing any where, it is most likely to be lying between Fort Garry and St. Pauls. I am sure Your active Young friend Kennicott will be well in Barney's District. The dangers of navigation and *John Frost* are what he has to fear most. McFarlane on Anderson's River is a clever Young Man and will be happy to meet Kennicott "A nunc congenial." G.B.

18. BERNARD ROSS to SPENCER BAIRD
Portage La Loche 6 Augt 1860
[SIA, HBC Corr Coll, Folder 36]

My dear Professor,

I am now in receipt of your very interesting and welcomed favors of 21 April and 2 May last and although I have really no time to answer them properly I must at least acknowledge their having come to hand.

I need not touch again on our collections, as I have already written you fully upon the subject, but I trust you will find the collection a complete one, and I again have to assure you that every thing in my power will be done to add to the efficiency of the mission. Mr. Kennicott will be treated with every kindness in my power and Endeavour to conduct the affairs as economically as possible.

I will devote my attention as heretofore particularly to the articles of Birds Eggs.

I scarcely think that the *Somateria v nigra* will again be found on Slave Lake.

Since you have so kindly offered to contribute the zoological articles to my intended work on the McKenzies River Dit. I must turn it out of hand so that by next Summer you may expect and have some of the proof sheets but as I will only get my photographic apparatus next year the views and portraits can go only then.

Your letter I will reply to next winter, meanwhile with best wishes to your family and yourself I remain, ever yours truly,

B.R. Ross

P.S. Along with this I forward a shrew please add it to my catalogue.

19. ROBERT KENNICOTT to LUCY HUNTER BAIRD
Mackenzies River above Ft Simpson August 21st 1860
[SIA, RU 7002, Box 37]

[Kennicott was fond of using native languages for effect, and he has done so in this letter. Although he usually used words from the Kutcha-Kutchin language that was spoken by the peoples living near Fort Yukon and on the Porcupine River, at least some of the words used in this letter are from the Co-yukon dialect. *Nazoon,* for example, is "good" in Co-yukon, while *Nīr-zīh'* is "good" in Kutcha-Kutcin. See vocabularies recorded in Frederick Whymper, *Travel and Adventure in the Territory of Alaska* (1869), pp. 343-50.]

Miss Lucy Hunter Baird

Dear Lucy—Lacoo! nenne tsakwe zazee nazon. Yes you *were* a good girl to think of the drawer full of Ledgers etc. to be sent in the box for me; and better still to write to me yourself.

I'm sitting in the stern sheets of a large barge which is going down La Grand Rivaine as fast as eight long oars in the hands of eight savages can propel it. I write here because I though you might fancy the letter the more for that, though 'tisn't the place to write neatly.

If you will look on the map you'll see a small Lake in the course of the river just below Slave Lake. We have just passed through it. It is not a broad lake as represented but some 20 or 30 miles long and only 5 to 8 in width. It is no favorite of mine for I had a hard time walking across it last March, and the musquetoes kept me awake all last night while on it.

You, like other "outsiders," seem to think that winter here must be terrible and summer the pleasantest season. It is true I like summer best, because I can shoot birds and get eggs and beetles, but I can assure you that so far as personal comfort is concerned winter is the most agreeable season. *Musquetoes* are far worse than 40° below zero. I have never been disgusted with the North till last month when I was almost eaten up every day.

I remained all last summer at Ft Resolution on Slave Lake, leaving a few days since when the boats returned from Portage La Loche. I will only be at Ft Simpson about a week and then I start for the Youcon with the hope to visit the Anderson River and perhaps Arctic Ocean ere I return to Ft Simpson two years hence. Then I'll tell you what kind of *Buttons* the Kutcha Kutchin and Eskimos wear.

I had pleasant letters from home, but was very sorry your father didn't get my letter from Ft Simpson and my heart is almost broken at learning that no arsenic and alcohol has come in for me. Ask your father if he couldn't weep in my place at such a terrible state of things. I think I should if it were not that I am a philosopher and smoke tobacco. But have had to smoke a great deal since I heard that I was to be so destitute.

When folk hear from home only about once in a year and then have to answer many letters so recieved in a few days while other things at the time distract their attention greatly, they are you will allow not in the best condition to write entertainingly. Therefore, such being my present situation, I trust you will pardon my sending you a very poor letter now, I intending to answer your letter properly next winter from the Youcon.

Please don't fail to present my compliments to Miss Julia Walter and Missie Jones.

I'm glad to hear you are getting on so well with your music and shall hope to be quite enchanted by your playing when I return. Please learn Beethovens spirit waltz. Roslin castle and some others of my favorites ere I come back two or three years hence. And meantime play something pretty for me occasionally of a Saturday evening.

Mr. Lockhart who is in charge at Ft Yucon sings and plays the violin very well and I hope to pass a much pleasanter winter than the last

Lucy Hunter Baird (1848-1913), circa 1870. Smithsonian Institution Photo #SA 263.

one.

I finish this on the 3rd of Sept a little above Good Hope and about 15 miles from Arctic circle, writing as before in the boat, which will account for this variation from my usually elegant handwriting. For our crew we have one Eskimo, one Canadian, two Youcon Louchioux, three Peels River Indians, one Hare Indian, the steersman is an Orkneyman, Mr. Gaudet's wife is a Cree halfbreed, and I'm a yankee—Pretty good variety!

Both Eskimos and Louchioux (or Kutcha Kutchin) are much finer fellows than the miserable Chepewayans and Slaves on the Upper Mackenzie. The day before yesterday we passed the outlet of Bear River, it is no *grand chose* of a stream. I expect to be on the Rocky Mts amongst snow and mountain goats in less than two weeks.

Please give my respects to Genl and Mrs. Churchill and kindest Regards to your mother. Yours *Hyperboreally,*

R. KENNICOTT

20. ROBERT KENNICOTT to SPENCER BAIRD

Mackenzies River - Two days below Ft Simpson, September 1st 1860
[SIA, RU 7215, Box 13]

[Dr. Philo Romayne Hoy (1816-92) was a physician from Racine, Wisconsin. He collected ornithological specimens for the Smithsonian, and he was also the "first to explore the deeper waters of the upper Great Lakes." (Dall, *Baird,* 1915, p. 285). Dr. James Graham Cooper (1830-1902) worked as a physician for two years after graduating from the College of Physicians of New York before signing on with the Pacific Railroad Survey Expedition as a surgeon. He then began collecting natural history specimens for the Smithsonian. After his work with the P.R.S.E. was finished, he went to Washington to work on his collections. He was, however, forced to return west due to poor health. Much of the remainder of his life was spent on the west coast where he continued to collect natural history specimens and write zoological articles. (Hume, *Ornithologists of the United States Army Medical Corps,* 1978, pp. 38-51). William Stimpson (1832-72) was one of Kennicott's closest friends. Between 1856 and 1865 Stimpson was responsible for the invertebrate collections at the Smithsonian, many of which he had collected while engaged by the U.S. North Pacific Surveying and Exploring expedition. In 1865 he was made director of the Chicago Academy of Sciences. The Academy had been revived three years earlier under Kennicott's direction.]

My Dear Prof Baird

I wrote you a hurried note at Ft Simpson just to let you know that, barring the non appearence of the preservatives, I am "in town with a pocketfull of rocks" as Dr. Hoy says when he means he is in a jolly situation. I was very busy during the few days we staid at Ft Simpson getting up my equipment and left my letters too long unwritten *as usual.* Mr. Ross had promised me an entire "piece" of common tobacco for the egging trade on the Youcon but when the posts were recieving their supplies twas fond the tobacco was 16 pieces (of 90 lbs) short of what came last year so I couldn't, of course, get it. But Mr. Ross being short of sugar I not only exchanged one of my three kegs for a bag of flour but another for tea, and with this tea I was enabled to get tobacco from various gentlemen and men, besides which I was permitted to buy a *Youcon* officers allowence of 20 lbs from the shop giving me in all some 75 lbs. With this and a large lot of the goods most in demand among the Yukon Indians I'll cause much wailing among the breeding birds on the Youcon next spring. I have cause to believe that birdnesting on the Youcon will go on at a much brisker rate and far larger scale than I could make it among the cursed swamp musquetoes and *Chepewayans* of Ft Resolution.

I have goods to pay Indians for crossing my pieces to La Pierres House so that I'll take them down at once this fall. Mr. Ross says I am not to pay for crossing my allowences and the payment for the other will not cost £5 with such goods as I've chosen. I think all I have taken from the shop this fall including the extra amount of tea, coffee and rasins which I was permitted to buy like the rest, and the bills given in payment for tobacco and other goods purchased of the officers and men, will amount to some £25 perhaps less. I reckon my Slave Lake expenses at £10. And I believe Mr. Ross does not charge me for the few things given for specimens at other posts. So I suppose I have some £20 on hand including the $200 you placed to my credit last spring, but not accounting for the cost of allowences and liquor from Red River or the cost of getting these in. Neither Mr. MacTavish nor the governor say anything to me about allowences, and Mr. Ross says they do not to him. Yet on the bills of lading from Red River are marked as being sent to me—2 bags flour, 2 kegs sugar, 15 lbs tea and 6 lbs chocolate (a clerks allowence). As Mr. Sinclair at Norway House has sent me two bags flour and 2 kegs sugar (he sent no tea) he stopped the flour and *one* keg of sugar so I recieved the full allowence sent from Red River with the addition of a third keg of sugar (but minus 7 1/2 lbs tea which was packed in a case that was left or lost). I don't know whether this is an allowence sent in by order of the "*Council*" free gratis for nothing or whether 'tis charged. At any rate I think I have you to thank for its coming. Mr. Ross says he thinks it not proper that I should be permitted to pay for this and says if 'tis charged to me he shall deduct the amt from the charges against me here. But come if dear or cheap the allowence is highly welcome and Lockhart, Jones, and I will live luxuriously next winter and will have one very large plum pudding and brandy sauce next New Years day when Mr. Gaudet is to *take a little run*! from Peels River down to the Youcon to spend New Years and Christmas with us. I gave away part of my tea to Clarke, Hardisty, and McKenzie with whom I was staying last winter and this summer.

The two gallons of grog sent me from home we drank at the Mess at Ft Simpson, of the 8 gals sent from Red River I gave Mr. Ross 2 of wine and divided the other six gallons among the gentlemen keeping for myself only the same share I gave each of the others.

I think they'll not forget it when the birds nesting season comes.

All the gentlemen promise to do their best especially at the birds eggs but 'twas very discouraging to have no arsenic and seed shot to give them. I gave part of the arsenic and shot I had to Messrs. Ross, Clarke, McKenzie and Hardisty, and alum and saltpetre to the rest, and Mr. Ross divided over 100 lbs of No. 3 shot among them, keeping 20 or 30 lbs for himself and giving me 15 or 20 lbs. Fortunately he had asked for two bags (of 80 lbs) of no 3 instead of one as usual. Mr. Mactavish sent me some corrosive sublimate and 8 lbs of alum and saltpetre, but the arsenic he sent proves to be only 3 or 4 oz, fortunately I had cached a few lbs of alum and saltpetre, about 3/4 lbs arsenic and a quart of pure alcohol, while the failure of the Slave Lake "fishery" left me nearly two gals of old, but still pretty good, rum mixed with alcohol. I distributed liquor for insects and Mr. Ross secured a quart or two of alcohol sent in for medicinal purposes. I also "stole" (I dare not own here that I took it) a quart or two of the Scotch whiskey. Mactavish sent me two gals of rum—*medicated fortunately*. He also sent in the seine and can I left with him four years ago and which had been sent without good result to Churchill River. He did not send all the rum I ordered mistaking the box that came from Chicago for one containing alcohol.

I left the seine can and rum of last year with Mr. Ross. I take with me about 1 3/4 or 2 gals rum, two quarts alcohol, a half pint ether, a few ozs of chloroform, 1/2 or 3/4 lb arsenic, 5 or 6 lbs alum and saltpetre, a little blotting paper for plants, some 12 lbs no 6 shot (I recieved 25 lbs no 6 shot from Mr. Mactavish also), 15 or 20 lbs no 3 shot, about 200 insect pins and 6 papers of pins from home, a little cork, a few insect bottles, etc. And my seine and two copper cans of course—one of the latter to be crossed to La Pierres House where I'll want to use the rum it contains for fishes etc. next fall. Not a very full outfit for certain but never say die! I'll bet my boots (and I've recieved a new pair from home) that I'll do far better next year than I did last.

Sir George Simpson writes me that I may stay two winters more choosing such posts as I prefer. Says I may go to the Russian posts if I like but at my own risk and expense. I of course will respect his request and use no strychnine for poisoning animals (but I've got two good fox traps). He writes that he "shall not fear any injury to the fur trade from my attempts at trapping." I take this to be in reply to my joking remark in my letter to him that I thought the amount of my hunts would not materially affect the "returns" of the District. But perhaps he means that I may keep all I can trap and this is what Mr. Ross thinks who tells me to skin all the fur animals I kill myself excepting silver foxes. Sir George says I may "obtain at the company's posts one complete set of the fur animals of the territory" but he says he can not extend the permission beyond this, and these are to be purchased of the company at the price charged to the companys officers which is £30 for a silver fox and so on.

I shall kill *myself* and skin a good silver fox and perhaps more than one each of the other foxes and wolves etc. if I can on the Youcon. And Gaudet at Peels River is to get a blue and white fox. But I'll only send out the one set from Ft Simpson if they are to be paid for and not any silver fox unless you say you can raise that am't, 30 pounds, for the special purpose. Of course if I don't have to pay for em I'll send em along. They will not be spoiled for furs so I can leave em at Ft Simpson if not to be had without too much expense. Last summer I told Mr. Mactavish of the three martins, two minks and the carcajou in the cases sent and told him to stop them if need be or that you would return them.

I'm very sorry my letter of last November did not reach you the more so as it contained the register of my specimens. But I trust it was only delayed as there was something wrong about my letter to The Grove. Mr. Mactavish writes that 'tis certainly not his fault as he pre-paid all of my letters. My package to you was a large roll of foolscap. Well sealed and properly directed I'm sure. Doubtless you have it long ago. But I will recopy the register sent then and those made since as soon as I reach Ft Youcon and will send them by the November Youcon packet which reaches Ft Simpson in time to go out by the spring packet. Hereafter I will be more thoughtful and always send duplicates.

You are really wonderfully clever. I find you told John Woodworth to send me nearly everything I asked for in the list I sent you in the missing letter. Only you gave too little arsenic. I would like to have 40 or 50 lbs come in next summer and have written home that a package of 20 lbs or so had best be sent seperately to Mr. Mactavish to insure its coming. I shall also write to Mr. Mactavish stating which of the boxes I wish to take precedence in case all of those sent last spring can't come. I shall ask him to send the box containing the arsenic first and if at least three or four cannot come to leave out of my allowence, if one is sent, the two kegs of sugar. I write in the boat going down to Peel's River. There are two boats, one with Youcon outfits, and besides myself as passengers Mr. McFarlane who is to take charge of the Anderson and Onion who is to be in charge of Good Hope, Gaudet of Peel's River and Lockhart of the Youcon. McFarlane does not however go to the Anderson till next spring. He goes up to Ft Simpson in December and returns to Good Hope in March. Onion is to go out to the Anderson to superintend the cutting of wood for the new fort which has to be rafted some "days" down the river to the proposed site of the Anderson. So the post will not be built till next summer. It will be actually on the borders of the barren grounds Mr. McFarlane says. Two or three miles east of the river begins the treeless region or true barren grounds, and a few miles below the fort there is no wood on the east side of the river at all.

The fort will be about 68°30' North Lat and perhaps about 120° to 125° west of Greenwich. About 5 or 6 days only *up stream* with canoes from the sea coast. To reach Ft Simpson in time for the Halkett boat in the fall I must leave the Anderson August 1st and walk 8 or 10 days across land to Good Hope. The water communication is poor and takes longer tho' a better route may be found ere I have to take it. *I shall go to Ft Anderson* if I can stay a fourth winter, and shall make my calculations accordingly. I will probably know next June whether I can do so but even if I don't get news till the boats arrive at Peels River in Sept, I could still if required go to Good Hope or Simpson by open water and to Athabasca on show shoes. But precious little doubt that I'll be permitted to stay. In which event, I conclude, after innumerable yarns with McFarlane, Lockhart, and others, I will come to La Pierres House in last of next August stay there and at Peels River till December when I'll go to Ft Good Hope with Gaudet to meet McFarlane who comes to have with Onion, Taylor from Ft Norman, Gaudet and myself a grand "blow out" on Christmas and New Years day. Thence I'll go to Ft Anderson with McFarlane, and after that to Halkett in fall; and in winter to Athabasca via Liard, Simpson, Rae and Resolution. Be in Washington the winter after, where I'll astonish the weak nerves of any unfortunate individual that invites me to dinner.

I did not see the Governors letter to Mr. Ross but McFarlane and Lockhart who did, have just told me that they understand him to say distinctly that I may keep all the fur animals that I kill myself, and as Mr. Ross told me the same I shall just go in for foxes and wolves next winter and if I'm wrong, why Mr. Ross is responsible. I'll act on honor though about the strychnine and kill nothing with poison. No martins at Ft Youcon but Lockhart and I will make a small hunt at some distance from the fort.

I don't like to approach the subject of my summers operations at Resolution after the boats left for the portage. It was just a dead failure. I only got a few insects and plants not a half dozen nests, scarce any fish and no mammals except poor "Funny" and "Droll" my young fishers and several rabbits. I made several long expeditions in vain and lost near a pound of blood daily from the musquetues in the swamps about the fort. But I'm "a philosopher wot smokes tobaccer" and shall not give it up as entirely a bad job so long as I have a knife for bird skinning, a bottle of alcohol, and a pipe to smoke. I did feel rather blue about it, but I conclude that even a little from this region is worth my staying for it even though I don't gain much credit or any at all by my collections. I begin to see that I can't expect to pile in the rare ones like my more fortunate fellows working in richer fields.

I really cannot think of anything of vital importance that I want sent me next fall in addition to what is coming. That is provided you have sent plenty of arsenic and some bird skinning instruments. I send home a small list of desiderata that need not come in till the fall of

1862. I would like to recieve a clerks allowence and 25 lbs extra tea each year and shall ask Mr. Mactavish to send it if you have not arranged about it. I have asked them to send me from Chicago several gallons of grog for next fall and about a "piece" for the fall of 1862. I shall order no more from Red River as I suspect it is pretty costly there and the *old whiskey* is preferred here. I think I'll ask Mr. Mactavish to send me two gallons of brandy next fall though, in case, *none* is sent from Chicago. In the fall of 1862 I can make arrangements to have what I want sent to Athabasca for my work in spring of 1863 there, but it would be well to direct a "piece" of preservitives and no 10 shot to me at Ft Chepewayan to be sent on to Norway House and be brought in *by the Athabasca brigade* which leaves Norway House late in July I think. If you find that *I can't stay a fourth winter*, better send me an equipment there next summer; It would be very well to send to Mr. R. Campbell, C.T. in charge of Athabasca District, at Ft Chepewayan some collecting materials anyway for his own use. If he don't use them himself I'll always find someone that will, I believe they have pretty good allowences at Athabasca so that I needn't have any for the little time I'll be there. At any rate I'll hear from Mr. Campbell on the subject as Mr. Hardisty has written to him for me.

I think I told you that only one box came in for me from the "states." It was one sent off from Chicago some days after all the rest and contained a couple of gallons of Scotch whiskey and some clothes etc. from home.

I wrote you by the summer boats asking for some light cheap printed cotton, large cheap cotton handkerchiefs, cheap bright silk ribbons, cheap pipes, etc. etc. for Indians. But not much *weight* need be sent. I shall order an entire "piece" of tobacco from Norway House or Red River. It will be the grand thing among the Eskimos, cost about 1 shilling stg. for the cheapest kind and will bring in more birds eggs than anything else.

I have some slight hopes that part of the pieces sent me may come in by the Athabasca brigade in which even Mr. Ross will send me down some things this fall. It seems that of the 10 boxes sent 8 were directed to Mr. Ross. I do not remember asking that anything should be directed to him beyond what was for himself like the books and sweetmeats. Hereafter better have all the things intended for my own use directed to me. Mr. Mactavish will probably charge the whole lot sent, to me, whereas several were really for Mr. Ross alone as books, sweetmeats etc.

Lockhart brought a specimen of *Jerabillus Labradorius* (I forget if thats the name, I mean the kangaroo mouse of Illinois and Wisconsin) from the Youcon and McFarlane brought *A. xanthognathus* from upper Anderson. I will enclose a shrew obtained by Mr. John Ried at Big Island. Lockhart says no end of Ducks, loons, waders, gulls, etc. nest about the Youcon.

I recieved by the boats your letters of Feb 13th (a long one) March 21 (enclosing Lucy's) April 16th enclosing Russian Letter and map; April 22 and May 2nd (the last my latest letter from home was May 4th. In one of your letters I found Coopers. In another Stimpsons—much obliged to both and to Ulke please tell 'em I'll answer from the Youcon, can't now.

I have just written Mr. Mactavish asking him to transfer from box no 4 the 10 lbs of arsenic, the corrosive sublimate and one bottle of chloroform to box no 6; removing from the latter wooden boxes and alcohol enough to make room to transfer from box no 1 insect pins, insect nets, cork, parchment, labels, and blowpipes and drills to box no 5 removing alum and saltpetre enough to make room; or packing in a larger box. To transfer From no 8 (in case it is not sent anyway) the bag of seed shot to no 9 and after these transfers to send the boxes as many as possible in the following order; first no 6, second no 5, third no 9, fourth no 8, fifth no 4, sixth no 3, seventh no 1, eighth no 2, ninth no 7. But I said in case you sent in time other boxes of collecting materials, these were to take precedence of the others after the three first. (no 6 and no 5 now contain about 7 gals alcohol, 28 lbs saltpetre and alum, a few 2 oz bottles, 25 lbs no 10 shot and 80 turned boxed for eggs, no 1 and 2 you know, no 4 contains arsenic, chloroform, ether, 24 4 oz bottles of alcohol, 36—8 oz do., no 3 is fitted entirely with paper for birds and plants. No 7 is filled with small bottles of alcohol. No 8 has 58 lbs sweetmeats for Mr. Ross, 25 lbs no 10 shot, small vials, pipes, some paper for plants. No 9 contains things for me from home: stationary, clothing, gun wads and caps, powder, paper for plants, strychnine, tobacco, *mouse traps* and no end of little things.)

I asked him in case he sent only five to make the transfers from no 1 and to try and send some of the paper from no 3 and then send in the above order. But I begged that if a seperate package of arsenic came, to be sure that was sent to me at any cost. I have told you all this that you might know in writing to Mr. Mactavish about sending in anything you've sent after the 10 boxes sent by Woodworth, and one of which (not numbered, and directed to me) I have recieved. I would feel pretty well off even with the two boxes and should be quite happy if five come including no 8. But I suppose the one you sent last to Mr. Ross (the eleventh) and no 8 will come in to him anyway and they might manage to send me the other 4—nos 6, 5, 9 and 4.

I order *nothing* this year from Red River excepting 25 lbs Louchong tea to be sent in *addition* to the clerks allowence, and I make no order for the latter, telling Mr. Mactavish that I suppose you have arranged it if any arrangement is needed (If tis sent me by the company why twill be all right of course). I order no grog *fearing* it may crowd out more important things. And as I got in so much this year twill be less matter if I've none to give em next. For myself I'd rather have the weight in arsenic, seed shot or ribbons, trinkets and calico to pay for eggs with.

Much more might be sent me by mail. Some of the gentlemen get in several lbs of newspapers by the *winter* packet. Please tell 'em at The Grove that I'm all right but can't write them from here as promised.

I've just written the governor told him I wanted to stay a fourth winter that I might visit the Anderson and that *[rest of sentence missing]*. Please send me newspapers by mail, not *strong* papers though. plenty books here.

[Signature missing]

21. BERNARD ROSS to SPENCER BAIRD

Fort Simpson McKenzies River 10 November 1860
[SIA, HBC Corr Coll, Folder 36]
[Captain William John Samuel Pullen (1813-87) commanded six naval expeditions on behalf of the British Admiralty between 1849 and 1853. In 1849-50 he wintered at Fort Simpson.]

My dear Professor,

Having now a sufficience of liesure I must reply to your two very kind letters dated the 21 April and 2 May 1860, which I notified you as having received at Portage La Loche.

None of the cases which you advertised me of, reached Red River at a date sufficiently early to take their passage to this District by the yearly craft: So that from want of sufficient materials I much fear that Mr. Kennicotts labours, will be materially interfered with for the present Season; but I do not think that there could be much difficulty in getting Mr. Kennicott's permission extended to another Season. I feel much obliged to you for your offer to mail me some newspapers, and will thankfully accept the same. From the Papers that you name I should prefer the *New York Times* and *Ledger*; but if these should be allotted otherwise, anything that you like, a *Republican* bias in politics being preferred, that is to say of the old Federalist party, although to tell you the truth I care very little for politics. They had better however be mailed directly to my own address, as they are then more certain of arriving in safety. Any Scientific works of which copies may be at your disposal will also be highly acceptable, and I will endeavor as far as lies in my power to forward the views of your Institution, as well as your own by making collections in all branches of Natural History and Ethnology as well as affording you any information on points of interest. I shall be most happy to give Mr. Kennicott all the assistance in my power. He is really a most indefatigable person, and has infected Several of our officers with his zeal.

I much fear that your anticipation of another *Som. v. nigra* being found on great Slave Lake will not be realized; but the officers in charge of Forts Rae, Resolution, and Big Island are on the look out, and if it be there it will certainly be secured, as well as the other desiderata you mention. The numerous cases which I forwarded to your address last Season, and the large collections which I am at present making will I hope prove, that my promises of aiding the scientific researches of your Institution.

Respecting the Eggs of Birds, I have been lucky enough to procure Several of your wantages, and hope in the Spring to secure you several more. I find it particularly difficult to make the Natives bring me the parent Bird, but in time they will probably do so. It would seem as if the old Jewish law of taking the Eggs and sparing the Mother, were a superstition among them. I have at present about 100 Eggs of different kinds well identified for you of my own collection: besides a number gathered by other persons. I fear however my stay in this District will not be of Sufficient length to complete the Series, and my Successor may not take so much interest in the matter as I do. I have now been in this Service from the age of 14, which as I am now 33 gives me nearly 19 years Service, of which period I have been 14 years in this District, *without any change*, and I naturally feel desirous of seeing a little of the world and taking a short respite from the cares of this immense District, which are not very light. I hope therefore, if I do not accompany next years collections to Washington, to drop in the ensuing Season, when I hope to have the pleasure of making your acquaintance personally, though from Mr. Kennicott's speaking so much about you I feel as if you were an old friend.

I am happy to say that I have a good number of Ethnological subjects for you. Such as Esquimx. dresses, arms and implements. Loucheux dresses from Peels River, and a complete Kutchin chiefs dress from Russian America. The latter is very valuable even from its cost, as it is very heavily garnished with beads, shells, and quill work, and cost many pounds to procure. I am not sure whether I shall retain the latter myself or not; but in any case it will be deposited in your institution until I leave this Service. I forward you a list of my Specimens so far as I have identified them but there are many yet awaiting their position. There are some rare birds among them.

By good luck I sent you the skin of quite a young Barren-ground cub last Summer, which I trust will prove interesting, and I hope to get you a Musk ox before long. Anything else you require speak freely and be quick as I may be off before I can get it done. I have the tusk of what I fancy to be an Elephant, brought from the Youcon River. It is so long that I do not know how to get it sent carefully, until I go myself. I have also some relics of Sir John Franklin including a gun and sword traded from the Esquimx. I will also direct attention to the fossil oxen.

I cannot really say whether the decline of furs depends on Seasons, or rather I should say on climatic differences, but I sent Prof Henry a series of observations carried over many years and dependent on the various writers of the Fort Journal. By comparing this with the list

Sent you can arrive at a very accurate conclusion the table of last year being enclosed to facilitate the comparison, and to add value to the result.

Fort Rae is situated on Great Slave Lake north. The Youcon at the juncture of the Rat and Pelly Rivers, and Anderson on the shores of the Inconnue, Beghooley or Anderson River. The latter is a Post which we are establishing for the Eskimo's trade. I enclosed a tracing with their positions thereon, as accurately as we can know without any Survey being made. In 1850 I was appointed to the duty of laying down the positions of all our Posts in this District, West of the Rocky Mts, and to fit myself for the same studied land surveying and the use of reflecting instruments, mapping etc. under Doctor Rae and Captain Pullen R.N., but Dr. Rae's proceeding on a searching expedition that year prevented my going as my services were required to carry on the work of the District in his absence, and the matter was afterwards dropped. The positions of these places are therefore only approximations.

The white-fox range is directly on the skirts of the Wooded country north of Great Slave Lake where those traded at Resolution are killed, though stress of weather has sometimes driven stragglers so far South as the Establishment itself. There is in my opinion the same difference between the white and the blue fox as there is between the silver and the red, but *no more.* I think that most of the marmots have been sent you.

It is perfectly impossible for me to say at present how many species of Canada Geese there are. Murray discribes a new on*[e]* as the *leucomelia,* and I sent another undiscribed, though not an uncommon one here, tis one which I called *Barnstonii.* I therefore think there are probably four, excluding the *Hutchins,* as for the number of tail feathers forming a distinctive mark, I cannot say that I would put much dependence on it, for my part, as the number of feathers vary from 14 to 18 in all kinds at least as respects size. The Blue wavy is found I understand only along Hudsons Bay, at least I have never seen it in here. McFarlane saw a specimen of the *Leucomelia* at Fort Good Hope.

I trust next spring to be able to fill up many gaps of the series of arctic waders and summerers, as well as to gather all kinds of Specimens which you may require. And I will collect notes for my work on the Ethnology of the District.

You will I trust accept my heartiest thanks for Your kindness in procuring the Books for me. They can however only reach me next Season. I hope that the German works are *translated* into either French or English as I will otherwise make but a poor hand at understanding them. The whiskey will also be welcome when it comes.

Mr. Kennicott has gone to the Youcon, where *I am confident* he will make large collections of great *interest.* And at all the Posts collectors are employed, I having taken the District from Fort Good Hope South, and he the remainder, so that between us we will sweep it pretty clean. Arsenic is however very scarce and other materials. The Egg business can however be carried on profitably, and as I do not intend to collect in future for any Institution but yours, reserving a set of the duplicates for myself, you may look out for a good number of Specimens. I have now enough for filling about *four* cases, from many parts of the District, and some collected in Summer on my voyage.

Would you be kind enough to purchase me a small silver Snuff Box at a cost of about $6, with the inscription To S. M. B. from B.R.R. engraved on the lid. The expense I will credit Mr. Kennicott with, upon hearing from you. If it could be safely transmitted by Post (registered letter) to the care of Burbank St. Pauls it would reach.

I was sorry to hear that Mr. Kennicotts set of notes were lost. A package of Ethnological notes, vocabularies, and discription of the Chipewyan Races, sent to Prof. Gibbs, New York, by myself were also lost. I will forward a set of notes on the aquatic and other Birds next Summer. Meanwhile, With best wishes, not forgetting me to Miss Lucy, I remain
Very truly & Sincerly Yours

BERNARD R. ROSS, C.T.

P.S. I was much gratified at your request thro Kennicott to have my photograph. I shall be most happy to give you a copy of the first that I get taken by which time I hope, by study and labor, to be more worthy of a place in your gallery of "illuminate" than I am at present.

22. WILLIAM MACTAVISH to SPENCER BAIRD

Fort Garry 23 Novr 1860
[SIA, HBC Corr Coll, Folder 33]

[James Robert Clare (1827-67), a Rupert's Lander born at Lower Fort Garry, joined the company as an apprentice clerk in 1845. He worked at York Factory for all but five of the twenty-two years he was employed by the company, and he was in charge of York Factory from 1858 to 1864. In 1864 he was transferred to Upper Fort Garry.]

My Dear Sir

I have duly received your letters of 8, 11 and 14 ultimo with their accompanying packets and several more besides. All will be sent on by Winter packet and will I think reach their destinations before the egg Drills will be required. Mr. Kennicotts packet will I think go safely enough and the rule will run no risk as all letters are carefully packed in packet Boxes.

I will forward your letter to the Gentleman in charge of Churchill to my friend Mr. Clare who has the charge of the District in which Churchill is situated, and beg him to lay injunctions on his subordinates to give the lemmings or Bear Mice no respite. All that I sent were from that quarter. In fact all the contents of the bottle sent you, with the exception of the small fishes, were from there, as well as the fur skins which were loose in paper in the box. I am very happy that there was anything in it worth while sending. I did not examine the contents as I have never dabbled in mammals. I knew Churchill to be a good locality, and if anyone was there who took a keen interest in the subject and was a little observant and active, he would make a great collection. When I first met Mr. Kennicott, it was there I recommended him to go, as it would take less time than going to McKenzie River. But the difficulty is with uninstructed collectors to get them to preserve specimens

Governor William Mactavish (1815-70). Hudson's Bay Company Archives, Provincial Archives of Manitoba.

of what they see every day. I know that many specimens of fish are got at Churchill which are not found in the country except far down the McKenzie and the Esquimaux sometimes bring in strange Trout. Is the Esquimau mouse as it is vulgarly called new? There was a young one I think among the Churchill specimens. They are sometimes found as far south as York. I will write Mr. Clare to get some more specimens of the White Partridge in spring when they are changing their color and in summer when fully changed. He is a collector himself and possibly may have some specimens by him. I one summer made a rather good collection of Eggs at York, but gave them away to an Egg fancier who visited us. Some of the eggs I got as far North as Repulse Bay.

I am quite surprised at the length of time the boxes from McKenzie River last year took to reach you, and hope those of this season will not be so long upon the way. Fortunately I got them this season just in time to send off by the steamboat, which was just going as the boat which brought your packages arrived at the Landing. The packages were shipped at once without examination.

Mr. Kennicott writes me that he got from a Gentleman, who had promised to collect a few plants for me in the Youcon, some of the specimens, duplicates which have been sent to you, or at least to the Smithsonian and request me to send more. I could not have done this before now, as the packet has only now come, but if you could let me know what he did send I would add as much as I can. I have not opened up the plants yet but there cannot be many as the packet is very small, so have no idea of what they consist of. I would cheerfully send the whole, but I am under a promise of long standing to send any collections of plants to an Edinburgh Botanist who is now in Canada. He has had nothing from me of late and writes me, reminding me of not exactly my promise but of what he has got from me.

Of course I shall be most happy to be of use to you or Mr. Kennicott, so do not hesitate to send his or Mr. Ross' to my care. If they only reach Pembina they will be duly cared for. I get all kinds of newspapers and magazines myself, more than I can always find time to read.

I have done nothing in the way of science unless I may call so my occupation for an hour or so past. I suspect you would call it prostituting science. It was simply trying to find out by examination with the microscope what a villanous compound palmed upon me by some of our Eastern Cousins under the name of "Louchong" was composed of. I have not yet been successful beyond that I can make out there are real tea leaves among it.

I think Sir Geo Simpson's death will not throw any difficulties in Mr. Kennicott's way. He purposes remaining another winter and no doubt he may if he chooses.

With much regard Believe me, Yours very truly

W. Mactavish

23. ROBERT KENNICOTT to LUCY HUNTER BAIRD

Ft Youkon, Youkon River, Russian? America Dec 18th 1860
[SIA, RU 7002, Box 37]
[Solomon Brown was Baird's assistant at the Smithsonian. He was also a leader in the nearby Afro-American community of Hillsdale.]

Dear Miss Lucy Hunter;

I certainly intended, as I promised, to try to write you an interesting letter from this queer corner of the world. But I fear I'll fail; not that this region is uninteresting, but because I'm too stupid to tell of it properly.

Do you feel interest in geography? Then let me say where "The Youkon" is, for as 'tis pretty certainly on Russian territory the H.H.B. Co would not find it politic to give the locality on the maps, at least it is *not* given. If you will get that same Arctic map of Arrowsmith's which your father has in the library you will see a small river laid down as rising in the Rocky mountains and passing La Pierres House (about 137° w of Greenwich and just north of Arctic circle). This, named Rat River, runs a little way westward and falls into what is called on the map the Porcupine River. This, which, here is called Rat River all the way from La Pierres Ho. to the Youkon, runs south west and empties into a very large and little known river at about 146° W long and 65° n, lat. (not exactly that). This large river is the Youkon, so called from its Indian name *Yu-k'hun*. k'hun is *river* but the meaning of the first syllable the Indians here cannot tell me.

On the north eastern bank of the Youkon some 4 miles above the mouth of Rat (or Porcupine) River is the fort. It is enclosed by high strong stockades with bastions at the four corners and looks something like a fort. Inside the buildings are placed on three sides of the square, leaving the middle and front open, as in the H.H.B. Co's forts I've seen. There are three buildings, a store, men's house and gentlemans house, all built of squared logs and roofed with bark of spruce trees. But the dwellings are tight, and with high fireplaces in which the wood is placed upright and burns famously; so we keep warm Indoors no matter how cold outside.

Often, in winter, the temperature is 50° or even 60° below zero but this has, thus far, been an unusually mild winter and yesterday, the coldest day yet, 'twas only 25° below. But strange to say 40° below is not found unbearable, and the men continue to work outdoors even when the temperature is 50° or 60° below, sometimes. Unless it be windy, voyageurs seldom stop for the coldest weather. Mr. Lockhart, the young gentleman in charge of this post, walked from here to Peels River and back during the coldest part of last winter. It is very remarkable how one comes to endure such cold.

Nor are you to suppose we go dressed in furs a la Dr. Kane. We are really more thinly clad in winter than I used to be in Illinois, and this from choice. We get plenty of moose and deer (Barren ground carabou) meat and when tis fresh and fat feast famously. We have also plenty of *tea* and each two bags of flour. Sometimes our cook regales us with "bangs" which are cakes made by mixing water with flour and frying the batter in moose or deer tallow. Better than mince pies or even oyster patties (is that the name of 'em) are "bangs"!—if eaten in the North at least. Sometimes too we get deer tongues, moose nose and beaver tails! When we go to the mountains hunting deer we have also deers *eyes*, marrow bones and occasionally porcupines. There is no end to our luxuries, though I fear you *outsiders* with uncultivated tastes would not appreciate them.

For firewood there is plenty of dry spruce and balsam poplar, which is hauled from the woods by dogs, and we've moose and deer tallow for candles.

Time never hangs heavily. Quite otherwise. I've had more than I could do since my arrival. 'Tis true we spend much time in smoking and talking; but that seems indispensible. Since the first snow, till within a few days when I commenced writing, I've walked from six to twenty miles every day, as have my companions, Mr. Lockhart and Jones, the officers at the Post. Lockhart and I were gone nearly a month on the mountains a hundred or two miles to the North west, hunting deer; the rest of the time we were trapping martins and foxes. The result of my fur hunt is second best at the fort among six older trappers; Of this I'm quite proud, and claim to be ranked above a *manger da lard.* One is much more respected here if a good voyageur, hunter, or trapper. I claim my rank only on the latter qualification.

Mr. Lockhart sings very well, and plays the violin, flute and accordion which three instruments he has, while Bras the postmaster has a jewsharp! We have grand concerts now and again.

Mr. Gaudet the gentleman in charge of Peel's River (Ft McPherson of the map) is to spend New Years day with us. He was to leave La Pierres House on the 10th and arrive here on the 24th of this month. Mr. Lockhart and Mr. Jones have gone to meet him on the road at a point five days from here. I was to have gone also, but staid to write letters, which letters I had like a naughty boy neglected too long while I was fur hunting. During the week that Mr. Gaudet stays no one will be permitted to write or do any work, so I had to let them go without me; though I'm terribly afraid Mr. Jones will get enough foxes to beat my "hunt."

You see people here don't care very much for cold or fatigue when one will walk four or five hundred miles and back for a visit of a week.

Next winter I hope to have a nice little walk in midwinter from La Pierres House or Peels River up to Good Hope and thence across to the Anderson; some five or six hundred miles in all; and all, except some 40 miles, within the Arctic circle. May be you'll think that I'll find it too cold. Not so. It is true I would be very glad to have it 40° below zero all the time provided it was never colder on the journey; but no fear we will suffer more in travilling than the feasts, pipes and fun at our campfires will make up for. Mr. Gaudet and I are to go Good Hope from Peels River, Mr. McFarlane from the Anderson, Taylor from Ft Norman and perhaps a missionary teacher from Bear Lake. We are all to meet about Christmas and stay till New Years day. Mr. Onion will need to keep all his dogs busy a month hauling meat before our arrival, or we will leave him to starve with an empty provision store when we go off again.

A droll Christmas party 'twill be don't you think? But this is a droll country.

When I return home I'll be able to tell many interesting things about life in the North that I cannot now. For so habituated have I become to the peculiar customs and institutions that they no longer strike me as droll or strange; but when I return to civilized society, if I am not to savage already ever to be admitted again, I will remember and notice these peculiarities. I am afraid I am by nature more savage than civilized however!

Even now I often have a laugh to myself at the queer style of our conversations, not only the topics but language. If I speak to anyone here, excepting the two other gentlemen, 'tis only about hunting, trapping, voyaging, dogs, or the Indians. Sometimes we get as far out as Ft Simpson, it being the head quarters of our little world, but we seldom trouble our heads about matters and places beyond them, at least in discussions.

The language mostly spoken by the men (servents) at the various posts is nominally French; but such French as would if heard by Miss Lucy Hunter Baird, cause her to make a grimace. (as you are studying French I conclude you do such things when *displeased in French.* Is it not necessary to make grimaces in speaking French?) The English is scarce better, from the number of queer forms and words introduced from Norwegien, Gaelic, Broad Scotch, Orkney, French, Cree and numerous other Indian tongues. I even observe some Eskimo words commonly used.

Please tell me where you spent last summer. What has become of your Uncle Tom Baird and Monsieur Dick Churchill?

Don't forget to learn some pretty music to play for me when I go back.

Please give my kindest regards to your mother, Mrs. Churchill and Gen Churchill. My respects to Mr. Haley is he still your clergyman? Does any body take good care of the snakes at the Smithsonian now? I wish I could be in it to night once. Won't you please tell Solomon to put a bunch of flowers on my old work table next fall. I suppose that is a very foolish request, but I think he will do it if he knows I wished it. All my memories of the old room are not connected with snakes!

Is your father's room at S.I. still the same? I wonder if there is anyone to come in now and again to trouble him with questions as I used to. Does *the* old coat! still exist? Yours truly,

R. KENNICOTT

24. ROBERT KENNICOTT to SPENCER BAIRD

Youkon, Youkon River Russian Am. Dec. 18th 1860
[SIA, RU 7215, Box 13]

[The native expression found in the addendum to this letter again appears to be a mixture of Kutcha-Kutchin and Co-yukon dialects. It may even be what Kennicott described as the "rubbaboo of a language we call broken Slavey, based on the tongue of Slave Lake Indians, mixed with French, English, Norwegian, Gaelic, Cree and even Eskimo words." ("Journal of Robert Kennicott," in James, *The First Exploration*, pp. 84-85.]

Dear Prof Baird,

I am going to try and astonish you by writing a short coherent and legible letter! If I do twill be the first you ever got from me; eh? I wrote you from Ft Simpson, Good Hope and Peel's river. I hired six savages and savagesses to carry my outfit across the mountains to La Pierres Ho and got most of it across; Lockhart, or rather Gaudet, with Mr. Ross' sanction, had my *allowences* crossed also. Apropos of your merciless strictures upon my misspelling of the word *gales,* conveyed in a hideous bon mot; I have to say that the Peels River "gals" *are* "strong" without mistake, for some small "wives" no larger than a southern girl of ten or twelve years old carried loads of fifty or sixty pounds across the mountains. As you suggest, the Eskimo ladies *are* stronger (in another sense) in summer than winter. I found them, men and women, far better looking and more intelligent than any northern Indians. But to be observed favorably it *must be from the windward*!—in summer at least. Heaven send me an attack of *influenza* during my summer on the Anderson!

I think I've forgotten the proposed coherency and run off the track already. *Wouldn't* I make smashes if I were a railway-engine driver?

We slept four nights in crossing the mountains. That is took five days or parts of them. Had a pretty good trip though 'twas snowy and we had to wade streams and marshes frequently. I carried no load and "took it easily" we arrived at La Pierres Ho Sept 18th (I think) and starting for the Youkon next day arrived here Sept 28th.

The Rocky Mountains where we crossed are no *grand chose*. They are bare of trees and covered with moss and lichens and mountain plants in places. Near the height of land there is scarce even a willow even in the valleys; but lower down there are patches of spruce and large willows in places along the small rivers. Cranberries, blueberries and some other berries grow everywhere and in a valley high up in the mountains we found red currants in greater abundance than I ever saw them before, growing wild. Some of the party saw sheep and deer, but I did not. The mountains are everywhere met; there seem to be marshes or springy spots even on their summits. I saw plenty of snow of the last year, in places, and it was snowing while we passed, so the aspect was decidedly arctic and chilly.

La Pierres Ho. is a "Provision Post", established, originally, to supply meat to Peel's River, as the deer (barren ground carabou) are hunted in the vicinity. It consists only of three small buildings. Here I saw blue fish for the first time. I skinned several and have since got more at the Youkon. At La Pierres Ho next fall I'll get, I hope and *trust*, a good many fish, mice, etc., and probably goats in plenty.

Coming down Rat (Porcupine of the maps) River I got a shrew, queer shrike, a hawk, which is either *Falco anatum* or the western sp, and some mice. Since my arrival here I've not done as well as I did at Ft Simpson last fall. I spent nearly a month in a trip to the mountains, about 150 miles n. west of the fort to hunt deer. I killed no deer, nor was I able to secure specimens of any killed by the others. But going out I trapped two fine foxes a cross and silver. We went with dogs to a small house built for a provision cache, at the edge of the mountains and 75 or 100 miles from the fort. From there we carried our packs, or got them carried by savages. Lockhart was obliged to remain some time at the small house after our return from the mountains, and I, unwilling to lose so much time from the fur hunting started off alone, to hunt foxes along the road to the fort. I finally walked in to the fort; took seven days to it, carried a heavy load, got short of provisions and tobacco, lost myself on the lakes, slept with one blanket when the mercury at the fort was 20° below zero. In short had quite a jolly voyage, or at least would have had but for the fact that not a fox or anything else visited my trap once on the voyage. But two nights after my return I caught a "cunning" silver fox that had previously cut his toe out of one of their traps and since been nightly tricking all the old trappers at the fort. This was a great triumph for a green hand. In fact I'm quite proud of my debut as a trapper. The first time I ever set a trap I got a fox, and in the two succeeding nights got one on each. I have two red, two cross, and two silver foxes, all fine specimens! My fox hunt (counting "skins") is second only to Lockhart's among six trappers. I have also 13 martins. I hope tis all right about sir Geo Simpson's consent to my keeping what fur I kill myself. I'll show you a splendid series of specimens of this fox, as well as the white ere I am done. I have two excellent steel traps that are as sure almost as a poison bait would be; so I don't much regret that I can't use the latter.

I have been quite crazy after the fur hunting, as we all have, since the first snow, we were, till within a few days, running night and day to our traps, and when we found no foxes within five or six miles of the fort we have made long voyages with dogs to reach better ground. We'll not get much now till spring. But then I must more than treble my hunt. I've a good martin road and the martins are thicker than usual. I shall trap for carcajous and wolves all winter when the foxes and martins don't run. Both the former are very numerous here but harder to trap than foxes.

I fear I'm given myself up too much to the trapping mania, as I've done very little else since the middle of October. The days are so short that it took the whole of daylight to attend to my traps properly, and we nearly always went out in the night to them too. This with a considerable amount of smoking, singing, and "*trapping talk*" has made the time fly only too fast for me. I hope to fill up my zoological register faster hereafter.

I shall be able to get a good many ptarmigans, woodpeckers, Hudsonian tits, cross bills and mice I think as soon as the days get longer. I have instructed the Indians to bring in porcupines, ermines, mice, shrews, ermines, marmots etc. etc. and, as you see by the register have already a few. I hope to secure a lot of the monster fossils found in this region and will have them sent to Ft Simpson to await your orders as I'm not sure about the propriety of sending such bulky articles. Would it do to saw one of the immense curved mammoths (?) tusks into sections for transportation?

I got as many fish as I could lay hands on last fall and a few mice and shrews, but I am ashamed to confess that I might probably have gotten a lot of ptarmigans and other winter birds by hunting for these after snow fell instead of the foxes and martins. I'll be a better boy next time! and try to make up for my naughtiness yet this winter, too.

The accounts I get of birds eggs here are most exciting. If, as I conclude, waders breed as abundantly here as ducks etc., I'll have a busy time in spring. I'll have about a hundred Indian boys at work nest hunting. There will be plenty of them at the fort. I will devote myself entirely to the egging in the season and twill be strange if I don't give a good account of myself. I will use all due care in the identification etc., and *you'll get no more pigeon hawks eggs without the parent* if I see her. I'll bear in mind all your hints and study the pamphlets which I recieved by mail. I have the blow pipe and drill. One I gave to Mr. Ross. Alas! I've no seed shot.

I cannot speak with certainty of many of the birds found here. *Colymbus Arcticus* is common in the fall and, they tell me, breeds here. Many gulls breed on the river islands. The banks are riddled with swallows nests, and I see more old nests in the woods than I have anywhere in winter except at Ft Liard. This is a land of lakes and rivers, not villianous swamps as at Slave lake, but fine lakes with high dry land between. I shall get a big canoe and employ a savage regularly to go with me till I learn to manage a canoe myself. The canoes here are exceedingly frail and easily upset and I'd do nothing alone at first. I judge from what you say that I'd better lose a rare bird or mammal than a rare egg and shall act accordingly. But I'll get both! only the egg *first*. I'll "make big eyes" at every locality like to contain nests of the waxwing and will have the Indians look sharp for these and other specialities.

I have one porcupine not skinned yet. It is not the one of the Eastern U.S.. 'Tis *epixanthus* I suppose. There is a Large blackish grey marmot on the mountains here and a small brown or grey one very abundant about the fort. There is another of the same size and perhaps a variety of the latter which is entirely black, not brown but deep black. I've seen several pieces of the fur. I'll get all these easily. I've obtained but one kind of shrew and one ermine as yet. I have two and perhaps three *arvicolae,* and I think *xanthognathus* is here also. I found the burrows and heard the peculiar cry of a colony of this species pretty high up on the Rocky Mts.

I never saw ravens more abundant than they are here even at this date. Hawk-owls are not uncommon and breed here. There is no *Hesperomys* in this region so far as I can learn. If there were one here it would pretty certainly be found in the stores etc. But *Jaculus labradorius* is here or at least some *Jaculus.* Lockhart brought one up to Ft Simpson last fall, which will be sent to you next summer.

The inconnui of the Youkon is I think a different fish from that of the Mackenzie. There is a loche here too. Blue fish are plenty in the small rivers. I think I have four species of *corregonus.* I don't know whether all the salmon are the same or not. They tell me there are two kinds. At La Pierres House I got what I think is a *cottus.* But I can give a better account of the fish when the water falls next summer. There are fine places for fishing in small rivers, sand bars etc. etc. I have three gallons or so of alcohol and Rum and shall keep dry skins of many specimens. There are I believe no trout here; though they say there are some in a certain lake far north in the mountains; and there are some very pretty spotted trout in a lake near La Pierres Ho where I'll get them. Those at La Pierres House are little over a foot long "grey with *red* spots" I'm told.

This has been a remarkably mild winter. The coldest day yet only 25° below zero. Last winter it fell to 55° minus and remained 50° below for a week. The coldest known here was 62° below I am told; yet it often falls to 60°. We will perhaps *get* it yet this winter tho it promises to be unusually mild as the time of extreme cold is nearly past.

My meteorological register has been very imperfectly kept on account of my hunting and trapping voyages. But I hope to keep it up properly during the ensuing time as I shall be away less from the fort. The old spirit thermometer that is here keeps exactly at the same point with the mercury one from the Smithsonian. So I shall send the latter to some southern post. When the new ones come I'll ask Mr. Ross to send the spirit thermometers (of which there will then be six) to Peels Riv, Good Hope, Anderson and Ft Rae, Leaving one at Youkon and Ft Simpson. The three mercury ones to Resolution, Liard, and Halkett. Next winter however I'll leave one at La Pierres Ho to compare Temperatures there and at Peels River during mild weather. Perhaps La Pierres Ho would be a better place than Good Hope (as the weather at the former place is often quite different from Peels River) for one of the spirit ther's. Will you not please ask Prof Henry and let Mr. Ross or myself know.

I am as eager and determined as ever to go to the Anderson I shall leave here next August I think with the boat. The boat makes two trips annually to La Pierres Ho. The first starts the first of June and returns about the last of same month, the second about Aug 10th returning last of September. The first boat brings down the outfit of goods. The second takes up the furs; both goods and furs being taken across the Mts by dogs in winter. The Gentleman in charge, going to Ft Simpson, leaves here in a canoe about July 4th, or 5 or 6 days after the boat arrives with the goods. Only two packets are sent and recieved annually, one sent usually in Nov (not till Jan this year) which goes out by the "spring packet" and Athabasa boats. The other in July, which leaves Ft Simpson by the winter express in Dec. One packet is recieved in the fall (containing letters from Red River by summer boats) the other comes down in the spring boats, and is the one leaving Red River in December. At the Anderson we'll get the winter packet from Red River in April.

I can, if I choose, go up with Lockhart in July to La Pierres Ho and after staying there and at Peels Riv over two months still reach Good Hope by open water. Or, if from any cause, I'm forced to leave for home (which Heaven forbid) I can go on to Ft Simpson and Slave Lake and thence out in winter.

But I think twill be better to go to La Pierres Ho in August, stay there and at Peels Riv till Nov or Dec and then walk up to Good Hope where I'll be at Christmas anyway. Gaudet wishes me to stay below till he goes up the last of Dec from Peels River. I'll leave the Anderson first of August about, and from Good Hope go in the boats to Ft Simpson. If I stay the 4th winter I'll probably go by the boats to Ft Halkett and then during winter walk out to Athabasca via Liard, Simpson, Big Island, Rae, and Resolution. I can if you wish go to Fond du lac on L. Athabasca or to Dunvegan on snow shoes. But Chepewayan ("Athabasca" tis always called here) is perhaps the best point. I'll there have the full egging season ere Mr. Ross' boats pass. (about July 15th)

But now for my latest scheme. Most of the rest you've had repeatedly to wade thro' already. It is to provide against the chance that I may not get the fourth winter here. I'm writing to the Governor to ask if I may go out in the early winter of 1862 in case he is not enabled to let me remain the 4th winter. Doubtless there would be no objections. I *must* go to the Anderson during the egging season so if I don't stay the 4th winter I'll leave there in August and go to Ft Simpson all the same; but from the latter point I'll take the boat to Ft Resolution and there hire a light canoe and two savages and carry on hard for the south. I'm not sure but that I can even reach Winnipeg by open water. If I learn that I cannot, I will stop at Isle a La Crosse and go thence on snow shoes via Green Lake, Carlton and Ft Pelly on the Assiniboine to Red River, a journey of about 40 days. This would cost me something more than going out by the boats; not very much however. Lockhart says I can reach Red River by open water.

I've given up all idea of trying the Russian posts, as the prospects are better elsewhere and I'm quite in the dark as to means of carring on operations there. Maybe within the next ten years we'll find ways and means to send me or some one else around by water!

It seems to me from all you say that the Anderson is *the* place for some of the rarer eggs. The post will be actually *in* the barren grounds and If the Eskimos are not tricky I can go to the coast, at least I'll get *[eggs?]* from them anyway. I'll bear in mind what you say about shells (mollusks) from Arctic ocean.

You mention that certain specimens of eggs etc. can be sold at high prices. If that be done to get more funds for the purposes of the expedition it is doubtless well enough. But if you mean any pecuniary benefit to accrue to me from the sale of my specimens, I object to it entirely. I never expected nor do I wish to recieve any pay for my services. All I want is to have my expenses (clothes included) paid, and then if I arrive at home clear of debt I'll be quite content with my pecuniary affairs. You see 'tis very pleasant to think myself an honest devotee to science. But I lose much of my own esteem if I work for money instead of science. And where there are none to esteem my efforts, the more need that I should in the highest manner myself.

Sometimes I am distressingly afraid that after all I may fail to make by my collections any adequate return for the trouble and expense of the expedition. But I trust such occasional fears are groundless; though I'm no longer expecting any such grand results as I have at one time anticipated.

Jan 2nd 1861

Lockhart informs me that in 1856 he went out from this district in fall by open water. Leaving Slave Lake in September (about 20th) after the return of the fall boat from Ft Simpson and reaching Red River in October. I could probably make the same time. Please give me your opinion as to the propriety of staying a fourth winter in preference to going out as I have proposed in fall of 1862. Unless Gov Simpson objects, you may consider me positively booked for a summer on the Anderson. The question is, if I get a fourth winter, whether the results of my fall at Halkett, winter at the various posts *en chemin* to Athabasca, and the spring, till July, at Athabasca, will pay for the additional expense. I think it will; but cannot tell how hard it may be to raise money, or even what the necessary clerk's "allowence" costs.

From what you say I infer that it would be best for me to stay the 4th winter. I am confident I could do much better at Athabasca than I did last summer at Slave Lake. It would doubtless be very pleasant to get home a year sooner. But I would ever after regret that I went so soon I believe. I do not now foresee any sufficient inducement for my remaining more than four winters; nor do I think I would like to stay longer, especially on account of my mother's wishes. I am afraid you will not advise me to stay long enough from fear that *I* may be the loser pecuniarily or have a hard time of it. But I assure you, I am glad to stay a fourth winter If I can thereby do any considerable good. I think myself that I would pretty certainly do more good by spending the fourth year here than I could at home. But, as I said, I am not sure you'll find it easy to raise the money, or that my collections of the last year will be worth much; and I would hardly enjoy myself enough to wish to stay for my own pleasure alone! At Athabasca I would have a full equipment and could work to much better advantage than I did last summer or can next summer.

I wish you would advise me to do just as you would yourself if in my place (and with only my small capacities for doing any good elsewhere, mind you). Then I'll be sure to go aright. But after all these rambling propositions, I fancy you, may consider me a nor west fixture till the summer of 1863. I will probably decide pretty certainly when I get your letters next spring, whatever. And you'll then know my decision when you get next winters packet in April 1862. If I don't hear from Sir George Simpson respecting the permission to stay a fourth winter, I will next spring write to Mr. Mactavish whether I will stay the 4th winter or not in case I do get permission, and he can thus tell whether to send in the clerks allowence. As you can tell him if I do get it.

You observe the packet leaving here in July goes out in the succeding December packet from Ft Simpson. We hope next spring to get the winter packet from Red River, here in May before the first boat leaves.

After December 1861 I will be able to send letters by all three packets. And will recieve yours within a couple of weeks after their arrival at Ft Simpson—Before their arrival there, in 1863.

I will buy a train of young dogs this winter (cost very little) which will be ready for work next fall and will be all I'll need for my voyages in winter of 1862-3.

I do not think you will need to send any more outfit for *my* use beyond what I've already written for. And I hardly expect to recieve near all I've asked for. Indeed I'm sorry that so much was sent to Red River last summer, as I fear some of the more necessary articles will be left. I've no idea that all will come in next year. I've written to ask Mr. Mactavish to send what I most want first. Still *everything* sent would be of great use if I could only get it. Even if I'm forced to go out in the fall of 1862 whatever comes in that year will be placed in the hands of others and bring good returns. But I *must not* go out till /63! Any light articles may be sent by mail and packets. In the fall of /62 I can send whatever I want there, to Athabasca, by the returning canoes which have brought down the fall packet to Slave Lake.

I trust you have found the missing letter I wrote you last winter in November from Ft Simpson but I've copied and now send you my register of that time.

Lockhart and Jones are very anxious to procure two of *Sharpes rifles* for the Deer and moose hunting etc. And I've promised to ask you to get them for them.

Neither they or I know anything of them except by hearsay, so if they are not like to prove useful in this country don't send them. We suppose they will at least throw a ball 100 yards with greater accuracy than a shot gun, that they can be loaded very much more rapidly, and that they are not so complicated and delicate as to get badly out of repair despite good usage. If they come up to these expectations, and do not cost (the rifle alone) over 40$ or $50 each, pray send two. The price will not be so much of a consideration as the usefulness of the article. I think they cost much less but, they want to make sure lest they should come at a higher figure than we suppose.

A rifle that will shoot straight at 100 yards or *over,* and can be loaded very rapidly is the desideratum, and Jones and Lockhart say if there is anything better for this than Sharpes rifle let the better one be sent. You or a sporting gunsmith will be the best judge. You must know that even with a single barrelled shot gun a hunter will often fire six or a dozen shots at a single herd of carabou and even at moose several shots are generally fired. While in the deep snow a good hunter will kill the whole of a brigade or 10 or 20 deer sometimes. From this you observe that a decently good hunter would with such a rifle as Sharpe's is represented to be, kill the deer *ad libitum*.

They (Jones and Lockhart) are exceedingly anxious to get these rifles and would consider it a great favor if you could see to their being sent. They promise to make great exertions among the birds eggs in return for such kindness on your part. Upon recieving notice of their being sent to Red River they will give me bills on Mr. Ross for the amt the rifles cost, and this will be placed to my credit. You can meantime pay the amt out of the exped. funds and charge it to me. I think if Krieder or some other good gunsmith would take pains to send the right thing and send priced lists of shooting apparatus they might get a large custom up here. Very many even of the men here have guns costing £10 to £22! *in England.* I think their English agents must humbug them in some way, for surely very good guns can be sold for less.

Any American gunsmith that could be relied upon would find it well worth his while to get a footing here. Tell Krieder, if he is the man to send them, to do the thing well and liberally and he may very likely get orders for a dozen as soon as these are seen here. I enclose a list of what is wanted with the rifles.

If I could have one this winter and next I would have grand fun, but 'twould be no use the third winter else I'd ask for one too and sell it again ere I went out.

I'll write you a long yarn about the Youkon during the winter, to go out by the packet next winter from Ft. Simpson.

For a week or so back, the temperature has been pretty uniform, always about 40° minus. I do not find this very cold! I go to my traps dressed as thinly except feet, hands and ears as I used to be at Washington! And for several days we've all hands played foot ball in our shirt sleeves, that is with only one flannel and one cotton shirt! Its very strange this ready endurance of cold. But mind you, all the time that it has been 40° below zero there has not once been the slightest breeze.

I'm afraid I've written a very unsatisfactory letter long as I've been at it. But tis *terribly* hard for me to write.

Please give my kindest regards to all the old megatheria and other friends in Washington, and make my excuses to Stimpson, Ulke and Cooper for my silence. Tell em I'll answer their letters next mail!!

How does John Woodworth come on?

If you write to Mr. Barnston please give him my respects and say I'll wait till I have been here longer and can tell him something of the plants and animals.

I learned with much pleasure from my father that your brother (which one?) was to spend last summer at The Grove. I wish Mrs Baird, yourself and Lucy had been going there too.

My kindest regards to Mrs Baird, Lucy and Gen and Mrs Churchill, and respects to Prof Henry and family. Yours always

R. KENNICOTT

Senne Yu-khun curra nazic—Senne diss Anderson manatta; Lacoo! Senne natha sa-koha manatla cooloo. That is so!

P.S. I write to Lucy by this packet, and send another envelope, like this, containing my register. I enclose a copy of my letter to the Governor that you may see what I've said to him.

It's possible that if I get good dogs I will go up to Ft Simpson from the Anderson in March 1862 and return in April. This would enable me to answer my letters and decide future arrangements on the spot (we would meet the packet at Ft Simpson). I only mention it as a possibility. I would have a very pleasant trip but I'll not go if there is a prospect of my collecting much in March and April.

25. BERNARD ROSS to SPENCER BAIRD

Fort Simpson 18 March 1861

[SIA, HBC Corr Coll, Folder 36]

[Ross refers to John James Audubon's *The Birds of America* (1840-44) and *The Viviparous Quadrupeds of North America* (1842-59). The latter study was written jointly with John Bachman.]

My dear Profr.

I was yesterday put in receipt of your welcome notes and letter dated respectively May 16, October 8, 11, 14. I should have wished that a detailed list of the contents of the various Boxes addressed to me for my own use had been sent, as if freight missed cannot be furnished, I may be obliged to leave some of them at Norway House. Hereafter will you also be kind enough to address *Mr Kennicott's* cases to him personally, as well as whatever you may send any other of the District Officers, merely addressing to me things for myself, as I shall have great difficulty in adjusting the charges for freight this season, all packages whether for Mr. Kennicott or myself having been paid on my account.

You will my dear Proffesr. accept my heartiest thanks for your kindness in procuring me the Ethnological works and also for the valuable and acceptable Books wh. you have so kindly and generously added to my private library, which for so high a northern latitude is something very respectable. Many thanks also for the whiskey in which I hope to drink your own and Mrs. Baird's health on Xmas day.

I much fear you speak too favorably of my exertions in collecting the curious objects of Natural History, not that I had not enough trouble with them on account of the want of preservatives, which occurs again this present Season, but you may rely that I will have *neither cost* nor labour to render the collection sent out by myself and brother Officers as one of the largest and most valuable contributions to Arctic Science ever made. It is now with me a labor of love, and I am daily becoming more and more attached to the study in particular of ornithology. Your books are in fact my only authorities: but none better can be required nor wished for. The completeness and accuracy *of your* discriptions are wonderful but I cannot say the same for the clearness of those of your assistants. The notices of *the Gulls* being in *particular very dim*.

I will as heretofore pay every attention to Mr. Kennicotts *comfort* and success. Endeavoring to manage his funds to the utmost and rendering him all the facilities in my power. He is a very zealous person and will render important aid to science. With respect to my own labors I am not getting as large a collection as last year. Many items such as squirrels, hamsters, mice, red polls, snow birds, etc. I am leaving out of notice as I suppose that you got enough last Season. My attention has been turned during the colder months to securing a set of the rarer arctic birds and the following have been secured, *Pinicola Canadensis*—male and female—*Picoides Hirsutus*—male and female—*Picus villosi*—numerous specimens—3 species of weasels—viz *[Putorius] noveboracensis, Longicauda and Richardsonii* in summer and winter dress, 2 Wolverines, 2 shrews, skulls of wolverine, Fox, Martens, Lynx, etc. *and a very fine female specimen of the Picoides Dorsalis*. These are all collected since I wrote to you, and are additional to the number I then sent you. When I have numbered the whole I think my *entire collection* will go to about no. 800 *at present* and to about 1000 when the Boats leave in June. But oh my dear Profr. Arsenic! I am afraid lots of specimens will spoil on my hands, and the *infernal* mice have already dished me 50 or 60, one of which was a Ger. Falcon I think. I have been most careful in my notes and measurements.

By the lamented death of Sir George Simpson, a fine specimen of the Musk ox Skull and Skeleton and one of the RM Goat, do. do. which I had caused to be procured for him, have become at my disposal. They will be forwarded now to the Smithsonian Institution, as well as the skin of another adult goat and young, wh. skulls, and the skin wh. hoofs and horns of a musk ox. Of the latter the 2d Musk ox, and one of the adult goats will be kept for myself to Deposit. The remainder are at your disposal. The new have been collected by Mr. McFarlane on the Anderson River, in whose name, that given the Smithsonian should be placed. He also gives me favourable encouragement respecting the procuring of a Specimen of the Skeleton and Skin of Barren Ground Bear. Among other articles wh. he brought up last fall were several specimens of a small marmot not as yet described I think with skeletons and Skulls, from the Anderson and other mountain streams. If not yet named I would suggest that they should be called *Arctomys McFarlandii* or McFarlanes Marmot. This gentleman also informs me that he has secured a *Picoides,* which is, from his description *evidently not yet discribed.* He represents that it has but 3 toes general color black, with *yellow head, back and breast.* I have not seen it myself, but will in Spring, when if undiscribed I should wish it called *Picoides Bairdii* after yourself. There are lots of other interesting specimens of all kinds too numerous to mention at present. Mr. Onion is making large collection of plants and Mr. McFarlane of fossils and other Geological matters. Any works *on the latter science* will be very acceptable to the last named gentleman, who collects with great energy. If you mailed a paper to his address I am sure that he would be pleased. My own are very valuable on *bird conds*, and I am therefore loath to give them away.

It is my intention to send you out a complete collection of the arts, manufactures, dresses, etc. of the Indians wh. copious notes. These are partly written in a popular form, and which you can make available for some scientific periodical sending me a few copies if they be worth publishing. As I should very much wish to become a corresponding member of some respectible American Scientific Society you could perhaps reduce them into a *thesis* which I believe is customary to present previous to election. I much fear that I am trespassing too much on your good nature, but I will do all in my power to return the labour. The Eskimo and Indian dresses are very handsome. A part is for yr. Institution and part for myself. The separation will be mentioned in the list when sent.

Many thanks for the paper of which a considerable number came to hand and a portion of wh. will after perusal be sent Mr. Kennicott.

I have a large Elephants tusk here from the Youcon. nearly certain—is it worth sending?

Should you have sent the N.H. Society of Montreal a set. *It was all is wanted from last years collection.* I will write you in summer respecting this years lot.

Pray secure me an octave set of Audubon's *Birds* and *Mammals.* Get the plainly and strongly half bound. Enclosed is a draft for £30 wh. will I trust cover all expenses. Pray see that they are packed in a soldered tin case within a wooden one, wh. freight paid to St. Paul.

The death of Sir George Simpson will not interfere in the least wh. you operations as respects this District. Within its bounds my authority is paramount, except special orders of Council be sent me on any subject wh. I am bound to obey.

I have now touched on all points which I need mention at present and in conclusion would desire to express the friendly feelings of esteem wh. I really feel for you though unknown. Such being not unusual in natives of the *Green* Island as I am.

Meanwhile wh. every good wish for your welfare and with kindest regards to Mrs. Baird and daughter in wh. my *better half* joins
I remain, My dear Professer, Very Sincerely Yours

B.R. Ross

P.S. *No case* but one for Mr. Kennicott containing clothing came on this year so that preservatives are much wanted.

26. BERNARD ROSS to SPENCER BAIRD

Fort Simpson, McKenzies River, 15 April 1861
[SIA, HBC Corr Coll, Folder 41]

[Mackenzie's *Travels* refers to the travel account written by Alexander Mackenzie entitled: *Voyages from Montreal, on the River St. Laurence, through the Continent of North America, to the Frozen and Pacific Oceans; in the Years, 1789 and 1793. With a Preliminary Account of the Rise, Progress, and Present State of the Fur Trade of That Country* (1801). Thomas Mayo Brewer (1814-80) was a physician, publisher, and businessman from Boston. He was also an avid ornithologist who collaborated with Baird and Robert Ridgway on *A History of North American Birds* (1874), as well as writing the pamphlet referred to by Ross, namely, *North American Oology*, Part I, *Raptores and Fissirotres*, Smithsonian *Contributions to Knowledge*, 89, (1857). John Cassin (1813-69) was an ornithologist from Pennsylvania. He was the Vice-President of the Philadelphia Academy of Natural Sciences. As curator of the Academy's collections, he was responsible for one of North America's most comprehensive and important ornithological collections. George Newbold Lawrence (1806-95) collaborated with Baird and Cassin on *The Birds of North America* (1860), but he later became an expert on the birds of Latin America. He was one of the founding members of the American Ornithologists Union.]

My dear Professor

I wrote you a few weeks ago, a letter which will reach you some months before this one will, and in it I gave you a faint sketch of my proceedings up to that date. Today, a few minutes ago, I finished Indexing my detailed catalogue, which as the latter will, this year, go in the last case, will give you an analysis of the contents of the Boxes. Accompanying each species are notes embodying such information as presented itself at once to my mind. As I am my own ammenuensis this winter, I was so pressed for time that I delayed writing these notices until the cases were packed, but so far as they go, they are correct, and must serve as an earnest of future contributions. I was unwilling, until I had studied the matter more fully, to enter upon the *Natatores,* but if I be blessed with health and leisure next winter, I will essay a paper for some scientific Journal. You must direct my attention to any point that you wish for particular information upon, and pray tell me if my Notes are properly written or not. The Collection now forwarded is the work of many Contributors, though I have certainly contributed by far the greatest part, and in any case this is merely the remains of last Summers collection, some that I gathered on the voyage, and the procurings of last autumn and winter. I have at present up to 800 running numbers and will I trust complete the 1000 by the 12 June, at which date I propose to make up the packages, any arriving afterwards must await Summer 1862. Can you tell me how many Specimens I have sent altogether including Messr. McKenzies and Reids contributions, which I am now sorry that I did not enter into my Catalogue as I have done this year. I also wish to have a corrected copy of my species, ie. the correct name of such as I have misnamed, which I fear are not a few. This must not be wondered at as I am entirely self taught and but a short time a student. I propose this year to keep type specimens to identify birds hereafter by, say a male and female of each species where I have a double set or more. Then after receiving a corrected list of species I shall be all right. This will also be a groundwork for a collection which I will hereafter make for myself. There are a good many Ethnological matters this year, some of which, as well as some Franklin relicks which I shall send, I wish to be kept in deposit for me until called for. There are a few articles in the cases for Sir W Logan, which you will oblige me by getting safely forwarded. And I wish a complete duplicate set of the smaller birds, mammals, insects and plants sent to George Barnston Esquire, care, Hudsons Bay Co. Lachine, Canada East. I also desire a complete set of all articles kept for myself, with this proviso that when there is but one article of a species it shall belong to the Smithsonian if they require it. The remainder is then entirely at your disposal. If you require any of the articles for your private collection take them. Any further consignments I shall inform you about and these remarks apply (except Mr. Barnstons) only to the list now sent. There are now 4 cases packed, but with the musk oxen when they come, and other collections, I expect my shipment of next Season (June 1861) to exceed the total bulk forwarded by both Kennicott and myself last summer. The articles on hand, in excess of No. 755, consist of models of winter dog sleds wh. harness and wrappers, Eskimos bows and arrows, specimens of *Gulo luscus (2), Otus wilsonii (1), AEgiothus linaria (2), Hesperomys myoides (7), Arvicola riparia (1), Plectrophanes Nivalis (3), Parus Hudsonicus (1).* The last I got today, it is a beautiful female specimen: but being rather pressed for time instead of skinning it, I merely extracted the entrails and threw it into alcohol. Respecting the preserving of

Ornithological specimens in this manner I wish to know if they are satisfactory as on the Voyage I might preserve numbers of Birds which I have then neither time nor inclination to skin. Is the extracting of the entrails an advantage?

Are you perfectly decided as to the *Parus* resembling the *Hudsonicus,* of these regions, of being that identified bird. The black under the chin appears to me a pure and deep black not a sooty tint, and as the bird which I obtained is a female the colors are of course duller than in the male. The size is also large for the genus the measurements being 5.40 — 7.80 — 2.55. Kennicott got two or three I think: but they are very rare.

Not one of the many Boxes sent came to hand, and the want of preservatives has been much against both Mr. Kennicott and myself. Cases to come in here should not be in Red River later than the 6 June. There is no dependence to be placed in the Athabasca Brigade whatsoever, as regards freight. I wrote Mr. Sinclair this Summer to send in every thing next July, at which time I hope to see the various interesting Books, and to drink your health and Mrs. Bairds in a glass of the kindly-gifted whiskey. Endeavour to send me the same quantity annually if you should be sending a case of etc. to my address. For want of the books I have not been able to proceed very far in my researches on the languages of the District. I am greatly in want of Sir Alexr. McKenzies *Travels,* to see the condition of the Indians when he first visited them, as I cannot bring matters to a satisfactory result, without it, I must write home for a copy though I am sure it is out of print. The Photographic apparatus is at York Factory: but when it will reach here I cannot say, and I am quite as uncertain whether I shall be able to use it or not when it does come. If I should acquire the use of it I will be able to take lots of portraits and views with it. Many thanks for the newspapers. *Harpers Weekly* is a very good one, and so is the *Ledger* though to call it a *news*paper, is rather a misnomer. The *Sandwitch Islander* is scarcely worth sending so far as it consists mainly of advertisements. I cannot quite coincide with Mr. Bonners opinion of his paper, for the literature in it, though sufficiently interesting, is far from *first rate.* In Mrs. Southworths tale of Rose Eliner, she mentions a Baronet of the creation of 1600, when in that rather anomalous order the earliest patents only be a date 1611, again fancy an English Peeress saying she was "*due* at Carleton House." Then the Poetry! with the exception of two or three pieces by Mr. Sigourney and a very beautiful one "The Tides" by Bryant, there is not a contribution that I think I could not compose myself. I suppose also that I have to thank you for sending me the prospectus for the new arctic voyage. I wish the affair success, though I have little faith in the open polar sea. At any rate, if they *do* stick, and strike for my domains I can assure them of a hearty welcome and every assistance in my power. Should they get caught let them skrike for the McKenzie or and soon as they get here. Once in they will be safe. The coppermine and Great Fish Rivers are bad of navigation and at a distance from our posts. For this prospectus it is mentioned that coffee is preferred to tea by Arctic voyagers, such is not the case among our people, who are as much exposed to severe cold as any men in the world. I fully expect when you get last years boxes that you will be pleased. For one season a pretty clear sweep was made I fancy. What were the greatest rarities sent? This year the musk ox will be the best I suppose. There is an Elephants tusk here from the Youcon—a *very* large one—which I will not send until I hear from you.

It was a pity that you did not get the tongues in proper time as the long journey must have spoiled them not a little. I shall endeavor to send you a few more. I am much gratified to have President Buchannan's autograph. Your list of desiderata will be completed as soon as possible. The Musk ox goes next Summer. The Barren Ground Bear and Raindeer will be procured as soon as possible, and although not mentioned, I am endeavoring to obtain *winter* mountable skins of Goat and Sheep. The Goats which I have packed up a few days ago are summer skins, but are good specimens, as they have the shank bones and skulls along with them. I shall devote as much time and attention as possible this spring to the procuring of Birds Eggs. I have already a few rarities packed, and hope to increase my list considerably before I close the cases. I trust that you will send me a copy of Brewers *oölogy* where completed if there should be any *to give away.* I think that after having been *murdered* by the moschiltoes in the way I have been and will be, that I shall have earned the recompense. This is an awful place for willows. They put one always in mind of Virgil's discription in the 3d. Book of the *AEniad.* "Et densis hastilibus horrida myrtus." and faith I do not know if Polydorus Iavalino pushed up into a more impenatrable thicket than that which surrounds Fort Simpson. The nests brought by the Savages have seldom anything more than *the eggs*, as to their identifying anything further than their own interest I do not put much faith in it. But I must pay them only for showing me the nest, or else when they snare the bird on it, as for shooting the parent, They would soon *create* a mother for the eggs, if they found it profitable. It is evident that they have not music in their souls, as they "are fit for treasons, strategems, and wiles and ought not to be trusted." Will you when you write enclose one a lot of rolls of Glycerine sticking plaister—it is first rate for strengthening fractured Eggs, also a case of preservatives, instruments, phials, cork lined boxes, insect pins, cotton wool, to my own address, after which I am bound to go in and win. I expect to send down *5000* specimens before I am done. Some pyroligneus acid is required.

The female snow birds don't become pure white like the males—*note* in the specimens *now* forwarded. They look like different birds, and indeed I named them separately last year.

I am endeavoring to obtain a specimen of the Strong-wood buffalo for you. It is much larger than the Plain variety, darker colored, and with very yellow Fat. They *seldom* come even to the little Buffalo River near Fort Resolution, and then not within some 20 or 30 miles of its mouth. Sir John Richardson mentions that it is found on the north side of Slave Lake at Slave Point, as well as that the red deer (Biche) is found in the Horn mountain country. Both these statements are errors. Le Biche is not found anywhere in the McKenzies River District.

By the by I wish that you would write a discription of the *Raptores* and *Laridae* yourself. Cassin and Lawrence don't discribe clearly at all. I feel quite nervous where I am obliged to refer to those portions of the book.

Many thanks for the two additional volumes of the *Pacific Railroad Report.* I hope you will be able to procure me vols 9 and 10 in particular. The 2 vols which I have are invaluable to me, indeed without them I should not take nearly so much interest as I do. It affords me much amusement to identify the birds as I procure them.

My number of collectors is pretty large now. Onion and Taylor have joined the band, and I expect to have Mr. Brass at Fort Halkett next winter who will make collections there for me.

In conclusion I will observe that every endeavor in my power shall be made to open up the natural History of these regions, both by forming collections myself, by my personal influence, and by affording every facility to Mr. Kennicott to pursue his investigations. And I trust that the consignments sent out from year to year will prove my promises to be borne out in fact.

Will you be kind enough to pay the cost of the carriage of Mr. Barnston's box to Canada, and if you will note me the cost I shall after place it to Kennicotts credit or refund it to yourself as you wish.

Remember me kindly to your little daughter. Tell her I shall send her a handsome pair of mocassins next Summer, which she must keep to dance in at her first ball. You will also please give her a pretty Eskimos needle case from the Boxes. I hope she may not find it too odoriferous.

I do not find that marriage interferes in the least with my scientific occupations. My better half is becoming quite interested in the thing and proposes making a collection of butterflies, but the pins we have are rather large for this purpose. As for *our parson* he does not like the business so well considering that when ever he visits me he is certain to get the whole list of my species, from memory, As I happen to recollect things remarkably well.

I had a letter from Kennicott (whose name here is *Joe*) and he is getting on famously. He has a set of fine Foxes for you; which is well as I cannot send fur animals without paying for them. There is no signs yet of the *Somateria v Nigra.* I am sorry that you do not possess the one that I was lucky enough to get. Have you read Murrays notes upon the Raindeer and Birds? They are very incorrect.

[remainder of letter missing]

27. ROBERT KENNICOTT to SPENCER BAIRD

Duck's egg Paradise! Ft Youkon North West Am. June 23rd 1861
[SIA, RU 7215, Box 13]

[Dr. Ferdinand Vandiveer Hayden (1829-87) was a geologist and palaeontologist. He participated in several western exploratory expeditions during the 1850s, and became a Professor of Geology at the University of Pennsylvania in 1865. His scientific career was disrupted somewhat during the Civil War when he served as an army surgeon. Antoine Hoole (d.1868) was the metis interpreter at Yukon, and he established the fort with Alexander Hunter Murray in 1847. Kennicott employs a native expression again in this letter. The distinctive diacritical marks indicate that it is of Kutcha-Kutchin origins.]

My Dear Prof Baird

Hurrah! for the Youkon every time! and hurrah for rare duck eggs (But D__n and double D__n all musquetoes in general and Youkon musquetoes in particular)

Lockhart and I have now something over a bushel of eggs of *Fulix affinis, F. marila, Aythya valisneria, Mareca americana* and *Dafila acuta* to say nothing of one nest of *Bucephala albeola*, one of *Nettion carolinensis*, one of *Spatula clypeata* and a solitary egg of *Melanetta velvetina*. I think you'll say we are good boys!

Lockhart has gone into the oölogy with what I've heard small boys call a "*regular rush*."

Since I wrote the above a *regular brick* of a savage brought me 76 eggs of *Fulix affinis* all found in the edges of one lake; and another (*pressed* brick like those Guy Cutting makes) brought Lockhart *Cynus buccinator* and 4 eggs. Iĕh lon-dĭ-ăs-tēnē illäh. See nos 1430 to 1437 in my register and Lockhart's no. 158. I now have off, hunting some distant lakes, my savage assistant with two assistant assistants, and Lockhart has Antoine and his other hunter each with their assistants male and female and we look for grand returns tonight or tomorrow. But with this style of things going on you'll not thank me to spend much time in writing what our registers tell.

I hybernated (mentally) as usual last winter. Began to *vegitate* when the birds arrived and became *animated* when I found the first good eggs, *Gallinago wilsonii*; since then I've been doing what Hayden calls *pitching in*, and can complain of no lethergy, mental or physical now. We got the packet which left Red River last Dec about the first of this month. In it were yours of May 16th and Oct 8th with two notes of Oct 9th and 10th—enclosures serpent paper, egg drills and pipe, needles, and Miss Bette's and Stim's letters.

If your letters to the other officers did one half the good the one to Lockhart did, you will have effected more for science by them than I shall in a years work. Lockhart was pretty well primed for zoölogical operations, but your letter "touched him off". And as quick as the spring boat was off for the outfit of the post he began and has been working not less eagerly than myself ever since. In fact I think neither of us could have well kept at it more steadily than we have. Antoine Hoole the (half breed) interpreter of the post, a keen hunter who knows the whole country, has been an invaluable assistant. I had learned from Mr. Hardisty his value as an assistant and have taken some pains to get him interested thoroughly in collecting from my first arrival here.

Deeming it of very great importance that Lockhart should keep up his zoölogical "enthuzy muzzy", which I believe will be best insured by his making good collections from the start, I made over to him as assistants Antoine Hoole and a savage whom I had hired and taught to collect, and I also have hired a savage "taxadermist" who skins the parents of our eggs. I have a good hunter engaged also for myself and lately engaged a second one. The payment to these five regular engagès (is that the way to spell it?) for the egging season will cost me but little; and I only wish I had had the organization of the egg hunt as complete from the first of the month. We get very little of value from the Indians, in oölogy at least, unless they are thus regularly engaged. Tho' both of us have obtained some good hauls lately from savages sent off to hunt a few days for eggs alone. We have only very lately learned the breeding grounds of the ducks properly and lost much valuable time hunting in bad ground. But all this smooths the way for Lockhart's operations next spring, when I predict he will do grandes choses.

The Indians too will then know what is wanted; and I shall not let any of them leave the fort this summer without recieving a long lecture from Antoine upon the immense importance to science of Lockharts recieving 5000 eggs of wax wing, *Picoides,* swan, Hawk owl and the like. I have made arrangements with Indians who spend the spring near the coast north of here to bring me eggs, and Jones with his boats crew have all promised to collect for me on the way to La Pierres House. I will however let Lockhart include Jones' collection with his. Jones and Lockhart both remain at this post another year. Whether both longer is not now known. Very likely Lockhart will leave next summer, and Jones be in charge thereafter at the Youkon.

I shall endeavor to get Jones interested in fishes and insects, shells, etc. this summer and it is proposed that he takes these departments, while Lockhart works up the mammals and birds, one taking the whole under his wing if the other leaves. I would suggest that you write to Jones (Strachan Jones) personally respecting fishes, insects, etc. and in view of his probably being in charge here after July 1st 1862. Jones is a graduate of Toronto College, a gentleman by birth and education and a *brick,* tho' what is called a dry stick. He puts me in mind of Cooper. He is the most industrious of the officers I have seen when he does go to work, and If he can be interested in zoölogy will be a valuable addition to the corps of collectors.

Antoine Hoole the interpreter of the post is as I have said a very keen hunter and takes kindly to the collecting, in which I have gotten him thoroughly interested; and he declares there shall be a very loud cry of bereavement among the parents of rare eggs every spring hereafter throughout this region. I consider his work and interest with the Indians a matter of prime importance to arctic zoölogical operations. I have bribed him with many very acceptable presents and shall give him some of the things sent from The Grove for my own use. That *accordion* you sent he has been very anxious to get, as he will. I have promised him that so long as he will collect well for the gentleman in charge here, I'll send him annually from the States, after my return, things which he is highly delighted in the expectation of.

I now know the wants of the gentlemen, as well as men and Indians here, so well that on my return I can be of great use to the latter as their agent in procuring certain parts of their supplies from the States. While I can at very trifling cost to myself repay the latter for their work in collecting in articles very highly acceptable to them.

For the egg hunt next spring I now furnish Lockhart with £5 to procure as many "skins" of goods as possible at Ft Simpson this fall, with which to pay assistants next spring. And I have given him 5 lbs of the tea that is coming, which he is to exchange (as he readily can) for tobacco. I propose that he should hire some 5 or six Indians for the month of June or during the full egg season. One or two will go the sea coast North or Northwest of here, who beside collecting will interest their companions who follow the carabou there. Another is to go far down the Youkon. Another far up the Youkon, perhaps another up Rat River. He will of course have several at work, as this spring, under his own and Antoines direction in the region of the fort.

I will this summer, with Antoine, tell every Indian who comes to the fort to bring certain special desiderata and with this organization and the experience gained by the collectors this spring—if good things are not "ground out," as you express it, I'll stop *turning the crank* for evermore.

Our registers tell their own story. I think you can rely on the species being as labelled. The species are few but the number of eggs of many rare ones large. I think you will be pleased with what we have done already, but I look more eagerly to Lockhart's and Jones' future operations. Remember the collecting season thus far has been devoted *solely* to egging. I pitch into plants and insects as soon as the egging season is over which I fear will be soon. The last half of July and all of August will show what can be done in fishes. Water too high before then.

I have four *arvicolae (good species)* one or two ermines, one *Lepus* (probably Ft Simpson one), two shrews, etc. vide register. What the *Arvicolae* are I cannot yet say, till I see the mammal book coming down in the boat. If any are new I would particularly like to see one named after Lockhart.

I hear good reports from the gentlemen in other quarters especially Mr. Ross, and Clarke seems to be *going into it* on the zoölogy. I hope you have written to them all. Do you not know that you have a persuasive way with you that enables you to make everybody do just as you like, as far as any external influence can have effect.

I think I can afford to go home—in the present state of zoölogical affairs here—after the summer on the Anderson and this I have concluded to do unless I hear from you this fall that there is any particular locality to be worked up here. The Columbia region and north could probably be best attacked from the west side. I will be ready for a descent upon any point of Nor West America west of Rocky Mts from Columbia to Bherings straits or Russian posts, if there are any, on shores of Arctic sea, within a year or two of my arrival at home.

The operations in zoölogy here are getting quite in earnest and we can now turn the crank and keep the "wheels" you mention going from the Smithsonian, with its long arms under your control.

There are various reasons why I deem it better to leave in fall of 1862 instead of wating till July 1863 as had been my intention till lately. Among these are the fact that you mention the probability of my getting little at Athabasca, of importance after the Anderson and Slave Lake have been worked up. This is true in most respects but *en passant* I hope to interest Mr. Campbell, and others in his district, whom I will then see at Resolution and we may get from them certain things in quantities (swans, both species, breed largely at Athabasca and Dunvegan is not to be neglected). Another thing to be observed is that the amt of my expenses for a fourth winter, placed in the hands of the various gentlemen here, in the shape of suitable goods for the egg trade, would produce perhaps more than I could with it. I too am getting seriously troubled by the effect of this monotonous life upon my mind and dread getting this lethergic habit fixed upon me. I was always lazy enough, but I used never to lose all mental energy for nearly a year at a time as has been the case here. My father says I may lay that to climatic influences. Partly so perhaps. Certainly 'tis only in warm weather I wake up.

Mem. Suppose I try South Africa or South America next time? There I would not be so long seperated from my beloved—snakes.

Therefore my movements will pretty certainly be as follows—(always supposing you do not chalk out any new road for some unforeseen exploration meantime)—leave here about August 15th for La Pierres Ho where I'll stay till last Oct. working fishes, small mammals, arctic birds and perhaps goats—Peels River early part of November; instruct Indians, Esquimaux, etc. and make arrangements there and at La Pierres house for the proper working up of the mountains. In last of November I'll harness dogs, put on my sled such of my more valuable winter collections as are not too bulky and heavy, hire a good savage to beat track, give my hardheaded *foregoer,* Nogah, (the carcajou) a preliminary flogging, tighten my belt and *tap* for the Anderson via Good Hope. I have four good dogs and the trip will not be a hard one. I shall very likely go with McFarlane on a spring trip with snow shoes along coast from mouth of Anderson to mouth of Mackenzie where he talks of making a geological exploration. But instead of this I may make a spring hunt for Barren ground animals, musk oxen, bears, hares, foxes, etc. to the east of the Anderson. If I can get McFarlane and his companion (two officers will be at Anderson) to pitch in the zoology fully at the fort, I will try to spend the egging season with a party of Esquimaux on the sea coast. In fact I will try the sea coast, if practicable, for certain. I'll bear in mind what you say of shells etc. I'll have to leave the sea coast by 20th July and Anderson by 27th to reach Good Hope by August 8th to 10th, when the Peels Riv boats pass for Ft Simpson. At Ft Resolution when I go in regular fall boat, I take canoe for Red River. I am asking Mr. Ross to try and secure me two voyageurs of those who would otherwise go out from district in summer, to take me to Red River. I shall have to pay them £6 to £8 each for the trip but this will perhaps be cheaper than hiring Indians, and much surer in many ways. I ask Mr. Mactavish to send me tea and flour etc. at Ft Simpson, Athabasca, and Isle a La Crosse; that I need not carry it all the way. I shall go as light as possible, and have little doubt of reaching Red River readily by open water.

Of course there is a long time yet before me, and circumstances may change my plans. But I think I'll carry out all the above. I take it for granted that no objection will be made by the company to my going out thus in the fall of 1862.

Mr. Ross asks me for Definite instructions respecting the division of the things sent in. I have told him to send me one third of all collecting materials out of which I will supply Peel's Riv, La Pierres Ho and Anderson (except in alcohol, of this I must have one third of the whole for myself alone because my Youkon and La Pierres Ho collections will require fresh alcohol, and I want plenty for the Anderson). The other two thirds to be divided amongst the other posts at Mr. Ross' discretion, only I have suggested that most of the alcohol should accompany the seines, and these (then will be two besides mine) be sent to two of what I consider the three best posts except Anderson, viz Rae, Liard and Youkon. If Clarke and Hardisty will use them properly better the two former, and I'll send Jones mine for use here in 1863. While the one from Rae can hereafter go to Anderson, the other to Halkett whenever there is a suitable person there to use it. I further suggested that so far as my present knowledge of operations goes I think Mr. Ross should take a double share of other things than alcohol, and that the Youkon, Liard and Rae should be as fully supplied as possible. But it is of course impossible for me to know who are making the best use of collecting materials. I shall divide the collecting materials the following fall myself unless you have indicated who is to recieve them, and thereafter we shall know pretty well what is wanted, and where it will *tell* best.

I told Mr. Ross to divide among all the collectors any *grog* that may come for *me*. The box of Goodies from The Grove were sent for him, and I of course give them to Mr. Ross. Some of the other things as music, pictures, etc. I have asked him to divide amongst the ladies and gentlemen and himself. Only asking for myself some plates of serpents! if any come. The other things sent me I direct to be sent to me mostly at Anderson, a few at Peels River. The spirit thermometers I suggested should be sent to Ft Rae, Anderson, Peels River and La Pierres House (there are now spirit thermometers here, at Good Hope, and Simpson). The two mercury ones will do for Resolution and Liard where tis seldom below 40° minus. I thought it would be well to observe variations in temperature between La Pierres Ho and Peels River where I will see that registers are kept. I have also asked for the barometer that I may try, if I can make out anything as to the height of La Pierres Ho, crest of Rocky Mts and other points above level of lower Mackenzie which is very near the level of arctic sea. I don't know enough of such matters to tell whether I can do any good thus but will try. If however you have told Mr. Ross to keep the barometer at Ft Simpson or any other special point it will not be sent to me. If it comes to me I shall carry it to Anderson and thence to Ft Simpson.

By spring packet I will write you about many things I forget now.

I will this fall go across to Peels' River for such things as I need at La Pierres House, and will probably have all the more important collections that are not too bulky crossed then and taken to Good Hope this fall in a canoe, to go out in summer of 1862. This will get the best parts of the Youkon collection home a year sooner than would otherwise be the case. Will cost £2 to £4.

I do not know what to say about sending in supplies for collectors. I have already asked for many things to come in in fall of 1862. All that comes I shall continue to make of use in one way or another; but small shot, arsenic and alcohol are of course the great wants. I wish you would, if you can, send Antoine Hoole, the interpreter here, egg drills and blow pipes. This post ought to have as good an outfit as possible, the Youkon zoological field will be almost inexhaustable.

If you have sent in alcohol, arsenic and small shot to come in 1862, with egg drills etc., and bird skinning instruments, some cork and pins for insects etc., I think perhaps there will be no good in my asking now for more. I however enclose a list of desiderata in case any of these should be omitted in what is already sent. On the other side of this list you will find a list of such things as will be best to send for the egg trade and I give the price at which these will be traded to the Indians (same as the companys Tariff for the lower posts). I have picked out the cheapest (in purchasing and dearest in skins) articles that are in great demand, excepting tobacco which is too heavy to send in now. But as tobacco is the very best thing for trade and is cheap I shall order (perhaps won't get it though) an entire roll, of about 90 lbs from Norway House to come in in 1862. I may require to repay this to the company for a roll I hope to get from Ft Simpson this fall, if not 'twill secure grand returns from Anderson, Youkon and Peels River in spring of 1863. It will cost me 1 shilling per lb perhaps less.

If these are not coming in (in fall of 1862) some of the articles I have mentioned for the trade in eggs, could you not send some in time. The weight would be very little and the matter is important. The gentlemen of course have no right to take companys goods to pay for specimens and at many posts, especially this, they can not even *purchase* them, then there being *too little* for the fur trade.

I sometimes fear that I am spending too much; but the field is large that we are working and to secure Indian assistance a good deal must be paid. I wish to see all the gentlemen at the different posts well supplied with means to keep the wheels going while *steam is up.*

I trust you will keep up a correspondence with all direct. Some would not like to go into this matter *through* Mr. Ross. I did not find out early enough all the peculiar little ideas and feelings of this little world. I fear I made a *faux pas* in asking that the labels should all be printed with Mr. Ross' name at top; this pleases him, but will not some of the others. Lockhart and Clarke will collect nothing for him. They don't see these things in their right light; I trust however that these little matters will not interfere with the general interest nearly all exhibit in zoölogical matters.

Mr. Ross was very kind last fall to me personally and gave me every facility in the way of getting my outfit of goods. We shall get on quite well now, and of course I fully appreciate his great exertions in the cause of science.

You will recieve from me (by spring packet) next summer, information as to who is going to be at the different posts etc. so that you can write fully, in time for the winter packet and egging season of 1863. Of course I shall do the best I can with my poor persuasive powers by letter, and next summer at Ft Simpson. The grog I hope will come. I wrote for a *piece* of good whiskey to my father long since, to come in 1862. Brandy is very costly, and not to be compared with the whiskey they all say.

Lockhart writes for books offering to pay for them and for collecting instruments. The idea of the bird or mammal book revised and with plates I got from once hearing you say something of its being republished by some book man or another. Of course the plates of *southern* birds wouldn't help Lockhart. But he wouldn't mind paying well for a book with plates or woodcuts of *northern* birds. If anything is out including northern fresh water fishes Better send me a copy to show, and to give to the best *fishermen.* Perhaps you can send something on insects. But anything on fishes or insects would require a glossary. Why don't somebody write some popular books on fishes, insects, etc.?

I have kept a meteorological register but havn't time to copy it. Will be more prompt next time. Jones promises to continue it.

I shall try and do something at Youkon Indian languages during summer with Antoine who speaks Kutcha Kutchin *better* than a native.

I'll send out some few dresses etc. I expect to be kept hard at work by fishes, insects and plants however.

I can get no good fossils as yet here. Have a piece of elephant tusk. Too heavy to send I think, but Lockhart will forward it if you wish. I'll of course keep a bright lookout, and will drum up the Indians about teeth and small fossil mammals this summer. On the Anderson I'll look sharp for fossils too and McFarlane is "pitching in" there. I hope to get some good things in fishes and fossils going up rat River this fall. No rocks near the fort.

Please tell Mrs. Baird that the enclosed letter is *purely platonic.* Very friendly though!, as was the one enclosed by you to me. That I was glad to get it you may be sure from my answering it at such a busy time. I'm writing only to the Folk at Home and yourself. I beg the megatheria to excuse me this time. I'll write by next packet to them. I beg Coopers Pardon for neglecting his letter so long. Am ashamed of myself. But I *can't* write when I'm hybernating. It comes a little tough to exist in such a *less than vegetable state* for 8 or 9 months for the sake of the 3 or 4 months of spring, summer and fall, a good part of which it takes to *thaw* one's brains out! But we'll see what my long voyage of something like 1000 miles in mid winter *within* the arctic circle will do for me. I'll *have* to look alive then. Mayhap you think I'll find it rather chilly. I don't know as to that. I shall drive my own dogs and have a savage to beat track and on the whole don't feel at all afraid of the voyage. *I've good dogs* and that is the greatest comfort to a voyageur. My dogs and harness, bells, etc. cost me £7. I shall readily sell them again when I leave for the same, or near that; perhaps more.

I really cannot give you any correct idea of how my account with HHB Co now stands for I don't know all the Red River and Norway House charges. I suspect Mr. Mactavish has not charged any freight on my allowences sent in last fall. I stand charged on Red River a/c books £9-3 in 1859 and £31 in 1860. At least these are the bills sent me. At Norway House £42 in 1859 and I suppose £5.5 for that keg sugar (tall price) in 1860. I really cannot say what my expenses have been in the way of shop accounts at Ft Simpson etc. as Mr. Ross sent me no bill this winter. As I reckon however I cannot have more than £200 at most charged against me from 1st 1859 to June 1st 1861, except for freight on allowences or on the boxes sent from home. Mr. Ross says there is a heavy freight bill against *him* for these boxes. I told him we would of course pay the freight unless he *preferred* to pay for the one or two boxes containing articles almost exclusively for himself. Next year (that is from August 61 to August 62) my expenses will be less I hope than last. The trip home need not be expensive; especially if Mr. Ross gives me some facilities which I doubt not he will.

I write this at midnight June 30 and just now saw the sun distinctly. We see the sun all night and have for two weeks past. This is by refraction I suppose. But I strongly suspect this post is farther north;

and west too, than is stated, very likely far into Russian America but we need not tell this as the company seem to wish it unknown.

Tell Stimpson please, that I get a little beastie very like a *gammarus* (is that it?) in some of the lakes. I'll look out for crustaceans here and on sea coast. I never saw anything like that *Apus* again. The spec was washed ashore on a sort of widening or lake in Mackenzies river. 'twas dead and a rather poor specimen I put it in alcohol with insects from Ft Simpson I think—Twas

[rest of letter missing]

28. JAMES LOCKHART to SPENCER BAIRD

Fort Youcon, Youcon River, N.W.A. 24 June 1861
[SIA, HBC Corr Coll, Folder 26]

Dear Sir,

Your esteemed favor of the 10th October reached me on 1st Inst. enclosing the Egg drill and blowpipe, for which accept my heartiest thanks. They have rendered me good service already.

You will have been surprised, no doubt, that no collection was made by me at the Youcon last year. The truth is this; when I arrived here autumn 1859, I began skinning, or rather attempting to skin birds and beasts, all that came in my way, but after a dozen failures, I gave up in despair and determined that I would do nothing until I could learn how to do it *properly*: for I was then aware that Mr. Kennicott would be wintering here with me, and would put me through, what he calls, "a course of sprouts." This he has very kindly done, and has taken an immense deal of trouble to drive into my thick head and clumsy fingers, the thousand little minutiae of collecting generally. It was useless for me to attempt to do anything during the winter, for Mr. Kennicott has gathered everything that was procurable, and had every Indian belonging to the Youcon engaged to furnish him with specimens. So soon, however, as the breeding season began, I, with my Interpreter, Antoine Hoole, began too. For all information regarding what I have collected thus far, I beg to refer you to a copy of my Register now forwarded.

It is my intention, if in health, to go in, and make a large collection of whatever I can lay hands on, that would be of Interest to the Smithsonian Institution; with the view, in the event of my going down to Canada to settle, a few years hence, of making a request to the Institution, for a small collection for myself; should this not be practicable, n'importe; they are welcome to all that I can do, to further the cause of Science. For this reason I have thought, that perhaps you would not deem it too presumptuous on my part, to request you to have procured and forwarded to my address, a few books and things to assist me, for without studying the matter as I go along, I should find collecting "flat, stale and unprofitable" work, in the extreme. Were there anyone else I know to whom I could apply for what I want, I would not think of troubling you, who have so much business to attend to; unfortunately, I know no one except a Father, brother and sister, from whom I have not had a letter for the last three years, and know not whether they be dead or alive. I enclose a List of the aforementioned "books and things." As soon as notified of the cost, I will give Mr. Kennicott a Bill for the amount; or should that gentleman have left this district, a Transfer will be passed to the credit of the Smithsonian Institution, in Red River, for the amount. It would perhaps be better to send the drills and blowpipes

James Lockhart (b. 1827) posing for photo while in the United States in 1867. Hudson's Bay Company Archives, Provincial Archives of Manitoba.

by mail, so as to insure their reaching me in time for use Spring 1863.

I trust in time to be enable to make an interesting collection, which I will send to the Smithsonian Instn. *myself,* in preference to transmitting it thro' a third party.

About July 4 I trust to have finished settling with all my savages for the summer. I will then leave here for Fort Simpson, and hope to pick up some interesting things along the route.

Accept my warmest thanks for the books you speak of sending me, they will be very acceptable, more especially, if they be anything bearing on zoology

I remain, Dear Sirs, Your's very respectfully

J. LOCKHART

29. BERNARD ROSS to SPENCER BAIRD

En Route to Portage La Loche 10 July 1861
[SIA, HBC Corr Coll, Folder 36]

My dear Professor,

Your consignment for this Season was completed 5 days ago at Fort Resolution, and consists I think of 15 cases, a Keg and Bundle. A considerable number of Specimens remained inland at Resolution, Big Island and Fort Simpson as they did not appear quite dried, and had no arsenic whatsoever upon them. Mr. McKenzie had about 105 Specimens including a good number of Eggs. Mr. Reid about the same number, and about 300 at Simpson including some of the Good Hope and Norman collections. I must hereafter, as the Birds and Mammals must be pretty well worked out, get the gentlemen to turn their attention more particularly than heretofore to fish, Eggs, and Insects, as well as fossils and plants.

The Shoes in one of the cases belong to Mr. Kennicott, and you will please find means to forward them to his friends.

The cost of the collections, Fort Rae excepted, has been defrayed out of my own pocket, and amounts both years to a considerable sum, upwards of £50, but I do not grudge either the money or the time, as I have learned from identifying the birds, far more than I could have done ever in a museum. I wish to make myself a name in the Scientific world if possible, and I am sure that you will do all in your power to gain it for me. I hope to get all your defficencies supplied eer I leave the District. One thing is certain you must send in a good supply of *necessaries,* including some "mountain dew" to be distributed among my followers. Try and send me a copy of Brewers *Oölogy,* with the authors autograph in it if possible.

Clarke has sent a large collection, and if they are well skinned they will be valuable. You had better mention those we can leave off collecting. I will conclude for the present by remaining Very truly yours,

BERNARD R. ROSS
Portage La Loche

My dear Professor. I received your very kind letters and will answer them by the winter packet. I have no time now. I cannot say what has come in as the packages are still on the road. I remain Ever yours truly

B R Ross

30. GEORGE BARNSTON to SPENCER BAIRD

Michipicoton 6th Nov/61

[SIA, HBC Corr Coll, Folder 2]

[Colin Rankin (1827-1921) left his home in New Brunswick for a position as Postmaster with the HBC in 1848. He worked for the company for fifty years, becoming a Chief Factor in 1879. Henry Youle Hind (1823-1908) was a geologist, instructor at Trinity College (Toronto), and editor of the *Canadian Journal of Industry, Science and Art*. He led the first Canadian sponsored exploration of the northwest in 1857-58, and wrote an account of his travels and discoveries entitled *Canadian Red River and Assiniboine and Saskatchewan Expeditions* (1860). Dr. James Hector (1834-1907) was the naturalist-geologist attached to the Palliser Expedition of 1857-60. In 1861 Hector was appointed geologist in New Zealand, and was subsequently made the director of the geological survey of New Zealand. Eugene Bourgeau was the botanist of the Palliser Expedition. Asa Gray (1810-88) is the botanist referred to by Barnston. Gray taught science and wrote *A Flora of North America* (1838-43) with John Torrey. In 1842 Gray was appointed professor of natural history at Harvard University. The Mr. Blackwell referred to here was likely Thomas E. Blackwell of Montreal. He was a member of the Natural History Society of Montreal, as well as being affiliated with the Grand Trunk Railroad. (Logan Papers, McGill University Archives, MG 2046; Minutes of Natural History Society of Montreal, 1958-1925, Blacker-Wood Library, McGill University). Elkanah Billings (1820-76) became the official palaeontologist of the Geological Survey in 1856. Billings had practiced law and edited the *Ottawa Citizen* prior to joining the survey. Thomas Sterry Hunt (1826-92) left the United States in the winter of 1846-47 to become chemist and mineralogist for the Geological Survey of Canada. He became Professor of Chemistry and Mineralogy at Laval University in 1856, and he taught at McGill University during the 1860s. He was also interested in museology and worked on behalf of the Séminaire de Québec before returning to the U.S. in the early 1870s.]

My Dear Sir,

I wrote to You from this on the 18th Sept, but have had only one opportunity of sending off the letter since, and that chance was lost through other matters fully occupying my mind at the time. I shall try to give you more news for this slip, and consign the neglected sheet to the flames. By next mail from the Sault which may turn up about Xmas, I shall be expecting the pleasure of Your report on the Case with the Horns, and the Skeleton of the new Otter, the "Destructor" about which I wrote You, when I booked it in the Express Office at Montreal. It would be very discouraging if either of these two packages had gone wrong and got lost. Mr. Rankin is also anxious to learn something of the minerals which You said the Chemist would look at when he returned in October to Washington. Mr. Rankin received the patent Office *Report on Agriculture* during my absence, and he desires me to express his acknowledgements through You, for its receipt, with which he felt much gratified. He now incloses for Mr. Henry two more months of the weather Kalendar, and he trusts they may be found interesting by the Commissioner of patents as well as by Mr. Henry himself.

I have now received a long Letter from Mr. Bernard R. Ross, who writes in good spirits. He had just received his Diploma of Membership from the Montl Nat Hist Socy, and was meditating giving them another paper for the *Naturalist.* It was to be an Ethnological subject, which has been called I believe the Science of Amateurs. Was this because the philosopher considered it as being the least exact of all the kinds of Knowledge man possesses of the Kingdom of Nature, and therefore most fitted for amateurs. Is it because Amateurs are fond of wading in Mist, and building Theories with an Infinitude of possibilities? My clever friend says he has got a new *Picoides,* which is a probability for there are far too few of these for northern Latitudes. The Spruces are able to support double the number of these Birds with the hosts of larva that burrow in them, and the Caterpillars of the *Cossi* in the poplars could afford them numberless Desserts.

I find by the list of Birds summering on the McKenzie that they have not many of our warblers. Is this actually the case do you think, or is there any sufficient reason for that family not proceeding so far north as other small birds? Bernard writes thus to me "I am doing my best to become a thorough ornithologist in addition to my other pursuits as an *Ethnologist.* My collections for this season comprise about 3000 articles, and will be sent to the Smithsonian Institution. I will however, write Dr. Baird to send You a complete duplicate set of the Insects, fossils, and Birds and smaller Mammalia, which You will please accept as a present from me, as I will defray all their cost to the Lachine House"

Now all this I hold to be very generous of Barnie. I must see for my part that he be no loser by paying the costs. And I depend upon You for the care to be taken of this present from the north, getting it packed carefully with snuff and camphor or other preservative and consigning it to my address, care of Hudson's Bay House place d'Armes Montreal, or John Swanston Esq No. 1 Windsor place, Dorchester Street, same city. Either will do, and I shall find all safe in spring there. It may be, *Deo volente,* and if all go well, that I shall visit the Smithsonian before two years have rolled over, when I shall enjoy the pleasure of thanking You, and having a regular good gaze at the beauties of Your collection.

I remember well the small shrew with the long tail, and I remarked this peculiarity to the children when it was first brought me. I forget whether it came from the pic or was caught at the House. The children have been hunting all October for mice and shrews, but have not succeeded in getting another, and I am anxious also for a *S. Palustris* for Montl Museum. My bottle contains 2 or 3 *Sorex Richardsoni, 3 Arvicola Gapperi, 1 Arvicola Riparia,* and *1 Arvicola* with red hair, perhaps *A. Gapperi,* but a variety having quite different color in the hair. The prairie mouse has been so plentiful this fall that a dozen has been killed in a forenoon close to the place.

I have not got any specimens of the *Salmo Ursina* yet, but have ordered them from the pic. They must be a quite distinct species. Every fisherman says they spawn in the fall, as it were, now the *Siscowet* not until February and March. I have had nothing sent me of late by friends, but Mr. McKenzie of Moose writes me that he has got a few Birds together again, and I hope they will be forwarded next spring. I got hold of Hinds *Report of western Explorations* in Montreal but the Palliser production by Doctor Hector was not to be had. If you can procure it for me, please forward it, and let me know the cost. The French Botanist that accompanied that Expedition was to have come out with something new, but I have heard nothing more of him since.

Mr. Kennicott I suppose to be in the mountains or across them for he has not written to me. I shall be most happy to hear of his welfare if you have any news from him. There cannot be a more indefatigable collector or one more zealous in the cause of science than our amiable Young friend. Let us hear something of him if he have written You.

I have been so busy this fall that I have had no time to ramble. The season was too late for Insects when I returned, and a few ferns and mosses was all I could pick up to add to my plant collection. All our Mosses here mostly are known in Europe if not to be found in Great Britian. I find Gray's descriptions a great assistance in the study of the Mosses. The more I see of the professors System and Ideas the more I admire his accomplishments as a Botanist.

I have so much to do, and so many long Letters of long standing to answer that it will be Xmas before I can begin to study, or get upon my hobby. I expect to have Your monograph upon the shrews when it is completed. Very little is known of the habits of these small Mammals. It is difficult getting at them.

The autumn here has been very mild and indeed we still have open weather without any appearance of the River freezing. I succeeded in having a pasture of flowers bordering our little Garden, quite a new thing at this place. If there be any thing new in the shape of flower Seeds or seeds of half hardy plants pop a few small parcels into the Box if You have not Yet despatched it. They will serve to remind me of You better than anything else, as I cultivate them.

Janet sends love and remembrance to the Dear Lassie Your daughter, and we all regret that when in Montreal we had so short a time of Your presence, that it was out of our power to shew You or Miss Baird the slightest attention. I sometimes regret that I left You when You went to call upon Mr. Blackwell, but I thought at the time You were seeing him on business, and feared Intrusion. The day after you left Montreal, Sir Wm Logan and Mr. Billings arrived, and before leaving I had the pleasure of seeing them both with Sterry Hunt at the Geological Rooms.

With best wishes for Your comfort and happiness in these sad and troublous times. I am My Dear Sir, Yours very truly

GEO BARNSTON

31. BERNARD ROSS to SPENCER BAIRD

Fort Simpson, McKenzies River 20 November 1861
[SIA, HBC Corr Coll, Folder 36]
[Bernard Rogan Ross married Christina, the daughter of Chief Factor Donald Ross, in 1860.]

My dear Professor,

I must now Endeavor to answer your two very kind favours of the 23 March and 4th April.

It is with great gratification that I learn that the Specimens sent by me proved of interest. The collection of last year will be doubtless as much so. That of next Season, will from leaving out so many species of birds, prove not so large but certainly of Equal interest as I have some *hundreds* of nests collected. I also rejoice to say that we have obtained on Great Slave Lake a *female* Eider, which from its size I consider that of the *S. v nigra*, though there be no dark mark under the chin, which probably is wanting in the Softer sex. The other rarities consists of the nest and Eggs of *Surnia ulula*, numbers of warblers nests, Ducks, etc. etc. and a good many Specimens of the *Colymbus adamsi*, skinned fishes from Bear Lake (no alcohol), etc. etc. I gave the copper tank with some alcohol to John Hope to collect Bear Lake fishes in. He will doubtless make a good haul there. He also informs me that the *Ampelis Garrulus* nests on Bear Lake in numbers, and has sent me several specimens of the bird. He says that the nests are very high up: but he will endeavour to obtain the Eggs next year. Next years consignements will be very valuable.

I see the mistakes that I made, some of which I had recognised before I received your list. I must confess I cannot see the difference between the *Empidonax pusillus* and *minimus*. There was also a bird resembling the female *Setophaga Ruticella*, but male—what is it?

I feel myself greatly indebted for your Kindness to the Montreal N Histy Society, of which I am a member, in forwarding the cases of Specimens. Will you be good enough to continue it to them from such duplicates as I may send out.

Sir John Richardson wrote to me informing me of his having presented the fossil bones to you. I have however a tusk and two teeth of *Elephan Americana* for him in return. There is also a tusk here which appears to me of far too small a curve for that of the fossil Elephant.

Respecting the *Bernicla Barnstonii*, I had a conversation with Mr. Lockhart on the Subject this autumn. He knows the Goose well, it is rarer here than on the Sea coast and he considers it *distinct* from the other Canada Geese. As to the no. of tails feathers making a difference, I think it quite imaginary. I am quite flattered at having the snow goose called after me. It is about the pleasantest compliment that could be paid me.

The cases of 1860 came in this Summer, but from not having been packed securely enough, greatly smashed, and many of the books damaged. The whiskey jars were all broken except 2, the contents of which were very nice, and I drank your health in a "moderate quencher."

Respecting Audubon, if you have not already purchased it, you need not do so, as I hope to get leave of absence next year, and it will be useless to bring it here to take out again. If it be purchased: but not sent off, it would be perhaps as well to dispose of it, though at a loss, which I will make up. None of this years cases came but I will get the books at RR "en passant" next autumn. Many thanks for them. The newspapers come regularly and proves very interesting. I think you will like the eggs of my blowing. They are seldom broken, except when the embreyos are too large. I shall send very few more arctic terns or chipping Sparrow Eggs, only some that are on hand. Should I go next year I trust that the zeal for Science will not slack in here. Push up *Lockhart* and *McFarlane* and *Onion*. The former of these gents got his promotion to the rank of C.T. this year. If you would write to John Hope, a schoolmaster of the Church Missy. Society at Bear Lake, he could get you some fine things. I have engaged him myself while I am here. Mr. Kennicott can procure you the Deer and Muskox in the Anderson where he is to pass this Winter. I have asked Mr. James Flett to procure you a *winter* goat skin.

Regarding the progress of meteorological observations I have addressed Profr Henry, as I hope (but am not sure) to see you next autumn. I need not write at great length at present, only assuring you that I shall do all in my power to send a *sublime* lot of things next year, including some native Skulls if I can get them.

I with kindest regards to Mrs. Baird, Miss Baird and yourself from myself and better half, I remain My dear Professor, Yours most truly

B.R. Ross

P.S. I will send a catalogue of what I have got collected next March.

32. GEORGE BARNSTON to SPENCER BAIRD

Michipicoton, 28th Jany 1862

[SIA, HBC Corr Coll, Folder 2]

[Georges Cuvier (1769-1832) was a French naturalist who revised zoological taxonomy by classifying animals on the basis of anatomical structure. He divided the animal kingdom into four phyla: vertebrata, mollusca, articulata and radiata. He is also known as the founder of modern vertebrate palaeontology. Dr. George Suckley (1830-69) went west in 1853 as an army surgeon on Governor Steven's exploration of the Washington Territory. He coauthored, with Cooper, the report on birds arising out of the expedition, and he contributed to the reports on mammals and fish. He resigned from the army in 1856, but served as Surgeon of Volunteers throughout the Civil War. (Hume, *Ornithologists of the United States Army Medical Corps*, 1978, pp. 420-42.)]

My Dear Sir,

Your favor with a copy of the Smithsonian *Report* for 1860 came to hand by our Winter Express, and your Communication of the 19th Oct with the list of the identified specimens of Birds, Nests and Eggs was also duly received.

The *Report* for 1860 is most interesting, and I am confident that all in our service who have helped to forward the objects of the Smithsonian Institution will be pleased to find that their assistance has been so highly appreciated. Mr. and Miss Blackwell must have passed a delightful week with You at Washington, for I believe both are ardent amateurs in matters pertaining to the science of Natural History. Your Kind Invitation to visit the Smithsonian I may one day gladly avail myself of, when the political Atmosphere becomes more peaceful, and the heated passions of men have calmed down to a temperate and friendly state.

Mr. Rankin has placed in my hands the Register of the Weather for the two last months, informing me at the same time that the Blanks No. 2 were all filled up. We have now no more of that number. If War should break out, I suppose there will be an interruption to all Communications, for the practice and Science of War are essentially obstructive.

I am happy that You find special characteristics in the Muffle palms and soles of the *Lutra Destructor.* There are scarcely any other exterior marks to distinguish it from the *L. Canadensis.* The size is less, and the Hair I believe is shorter, but then there are Young otters of the species "C" which are small also, and the length of the Hair may vary according to the Season. The Cranium is the surest criterion in this case to judge by. If there be no sufficient details in print of the *Lutra latucines* of Cuvier to identify it, I think the specific name *Destructor,* derived from the Indian, and descriptive of the Habits of the animal ought to be retained.

I remember well the "small Mole" I sent You, and feel happy that it has given an addition to Your own Genus *Neosorex.* I cannot recollect *when* I despatched the other specimen, for want of time for examination often forces me to throw these sorts of things into the Boxes and vials without notes. It surely could not have been the first long tailed shrew that I sent supposed to be the *Sorex palustris.*

Like Dr. Suckley I am not quite certain of the difference between the *Salmo Siscowet* and *Salmo Macqua* say *Ursinus* being specific, but I am disposed to think so, from their having different Seasons for spawning and the Indians recognizing the two as distinct fish. I must say also that I find a difference in their shape and the distance of the Eye from the Snout, and the Top or rather plane of the Head above, that part being flatter in the *Siscowet.* In other respects externally they resemble but the body of the *Siscowet* is rounder transversely, and not of such depth as in the *Macqua.*

Hoping that You and Your family may be preserved from the Evils of War, and that God in his Love, and by the power of his spirit, may guide the Hearts of all to peaceful Issues. I remain My Dear Sir, Very sincerely Yours,

GEO BARNSTON

33. WILLIAM MACTAVISH to SPENCER BAIRD

Fort Garry 26 Feby 1862

[SIA, HBC Corr Coll, Folder 33]

[Chief Factor William Joseph Christie (c.1820-99) was in charge of the Saskatchewan District at Edmonton, 1858-70. He entered the service in 1843 as an apprentice clerk, becoming a Chief Trader in 1854, a Chief Factor in 1860, and an Inspecting Chief Factor in 1872. The Christie name was well known in Rupert's Land; William was the son of Chief Factor Alexander Christie, who was the governor of Assiniboia during the free trade disputes of the 1840s.

The Institute of Rupert's Land was established on 12 February 1862. Governor Mactavish and Dr. John Schultz were elected secretaries of the Institute, while the Anglican clergy assumed the Presidency and all but one of the Vice-Presidencies. The Council consisted of the leading citizens of Red River, including William Caldwell, the Editor of *The Nor'Wester,* and Donald Gunn. Although the Institute only met twice, there were grand plans laid at the inaugural meeting held at the Court House. Its members planned to collect scientific data and specimens, to establish a museum and library, and to exchange their collections with other societies. It was also suggested that a telescope be purchased in memory of George Simpson, and that a microscope be bought in honour of Dr. Bunn. (T.C.B. Boon, "The Institute of Rupert's Land and Bishop David Anderson," *Transactions* Historical and Scientific Society of Manitoba, Ser. III, No. 18, 1960-61, pp. 92-111.)]

My Dear Sir

The Northern Mail came in a few days ago bringing letters from Mr. Kennicott. Your letters by it are forwarded now. Mr. Kennicott tells me he means to come out autumn 1862 and expects to be here in November and requests me to send to different posts en route supplies for his voyage out. I suppose if you desire it he will proceed to Churchill instead of going home. If it is intended that he should to there he will require no supplies for his winter from here as he can get them more advantageously at York. But I would be glad to know your intentions in the matter as it would be better to send his supplies to Churchill in July and he will then travel light and can go from York Factory to Church on foot late in the autumn. If, however, he only reached the Factory in November he would require to remain there till January as from drifting ice in the River they are impossible until all is set fast.

Mess Christie and Clare are now here and both have promised to interest themselves in getting specimens for you. They tell me that the Governor and Committee have issued orders that all specimens in Natural History collected for Individuals or Societies *in Britain* are to be sent to the Hudsons Bay House and they will be presented by the Governor and Committee. I do not know whether the in Britain is intended as an exception in your favor, but it is not improbable that it is as I know that Professor Henry's handsome letter was gratefully received by the Governor and Committee. I am told the reason for the new order is that a Mr. Murray, who applied to the Company for specimens from the Country after getting a great many, presented them to the Kensington Museum without a word of acknowledgement to the Company. I fancy the unfortunate acted unwittingly, and am not sure that the Company deserved much credit as the officers at their own expense and trouble got the specimens. But still Mr. Murray deserved some punishment for his want of savoir faire, and I am much mistaken if he will even get the specimen of ***Lagopus Leceuras*** that he is so intent on. Curious that such a man should not be better informed as to the habitat of the bird, but he is searching for it on the shores of Hudson's Bay. I will see that the new order does not affect your interest as you do not come within the restriction, but if opportunity offers I would recommend a complimentary and grateful letter to the Govr. and Committee. Perhaps Mr. Kennicott's return may afford an opportunity. This is of course entre nous.

The good folks here have started a scientific society here to be called The Institute of Ruperts Land. I fancy I see you smile, but if the Society hang together it will have specimens from the Country to dispose of, and for it I am in duty bound to beg your contenance and support as I am appointed corresponding Secretary, and as such will have the pleasure of forwarding to you a prospectus and report of the first meeting. I may probably also place several copies at your disposal and beg the favor of your distributing them as may appear most advantageous to the interests of the young Society. I count upon your support and will do all in my power to secure for you spare specimens. Our funds are as yet not very large, but subscriptions for a Telescope and Microscope are pouring in at an astonishing rate. It is proposed to purchase first rate Instruments (£150 and £60). I tried to moderate the furor and hinted that a good working microscope could be got for £15 and that the balance might be applied to other wants, but I was fairly swamped *[blotted]* the more effectually to shut me up asked what my own microscope cost. I could say no more, though I knew that none of my assailants could set up the microscope for use when it came out. It will do good however and a few will continue to use the Instrument and a great number will now take to collecting which is an end to be desired. Believe me to be, Yours very faithfully

W. MACTAVISH

P.S. I see by an order from the U.S. Post Office department that the postage on letters from the States for this Settlement is to be 10 cents for a single letter. This is unjust as we get our letters from the Pembina Post Office. If you direct your letters to me at Pembina they will reach me as certainly as if addressed to Fort Garry, and the three cents will cover a letter. Mr. Kennicott's letters can be addressed in the same way.W.M.

34. BERNARD ROSS to SPENCER BAIRD

Fort Simpson McKenzies River 1 June 1862

[SIA, HBC Corr Coll, Folder 36]

[Albert (Abraham Alfonse) Gallatin (1761-1849) was an influential Pennsylvania politician and banker. He was also an important member of the early nineteenth century scientific and intellectual community. He was the founder of the American Ethnological Society (1842), and a President of the New York Historical Society (1843). One of his most important scientific contributions was a *Map of the Indian Tribes of North America* (1836), and it was this work that influenced Ross' study of the Dene. Ross' observations on the Dene were published in the Smithsonian Institution *Annual Report* for 1866, but the manuscript he submitted was edited to exclude his discussion of origins, migratory patterns, tribal divisions, population estimates, and Dene linguistics. Compare "Notes on the Tinneh or Chipewyan Indians of British and Russian America," *SIAR*, pp. 304-11 with B.R. Ross Notebook, SIA, RU 7221.]

My dear Professor,

I have now to acknowledge the receipt of your two very kind letters of the 2d and 6 of Novr., which I must now reply to as our short summer, and its numerous duties are at hand. It is a pity that the cases had not all arrived before you wrote, as I should then have got an account of the value of the collections from you: but it is a pleasure I hope yet to come. The Register was put into the last case to save postage: but as there seems to be rather a reason for sending it seperate, I shall do so hereafter. Were my time not very fully occupied I would copy it in duplicate. From the nos. which you mention as having arrived some of my best things were yet behind.

The collections of the present Season, up to the 26 Feby have all been carefully packed in 6 cases, and there are enough for 2 more yet on hand independent of what my Sub-collectors will send in—in all I suppose that you may look for 10 cases from Ross and Co, exclusive of Kennicotts, Lockharts and Clarkes. In the Egg line I think that you will be set up this year, I have already packed 260 nests, and I hope for a few more, though the Egging season is just the time when I am most busy, closing up the yearly accounts and preparing for my Summer departure. The Recapitulation of the collections to the 20 Feby is also Enclosed and amounts to 704 articles. It will be hard if I do not make up the 1000 before I start.

The collections at Resolution, Big Island, Lapierre's House, Salt River, and this place have been made at my Expense. Though I give credit to the Gentlemen in charge of those Posts for procuring them.

Among the articles you will find a goodly number of *Picoides Hirsutus, Larus Hudsonicus, Pediocetes urophasianellus, Lagopus Albus, Charadrius virginicus, Columbus Adamsi* and other of your wantages. Besides I hope to get several more of the *C. Adamsii etc. arcticus* when I pass the Big Island. The former of these is found almost entirely on Great Slave Lake.

With respect to my own portion of the collection or rather what I wish disposed of for me. I wish a complete *duplicate* set of Mammals, Birds, Eggs and insects sent in my name to the Montreal Natural History Society and particularly specimens of the *Grus Americanus, Colymbus adamsi*, and other rarities. There are a male and a female sheep with skeletons in the cases. If not positively required for the S.I. would you be good enough to send *one* of them to this same society.

I was *highly* gratified at receiving the notification of being elected a corresponding member of the A.N.S. Philad. In the cases are two Kootchin skulls for that Institution, which I wish you to send in my name: but Mr. Gibbs may have the use of them previous to their transmission, any other article which you can spare in the cases and which you think would be acceptable you can send also for me. I hope to get a musk ox skeleton for it from Good Hope, with these exceptions the remainder are at your sole disposal. If not too much to ask you might continue *my own* series of Eggs and keep a set of the birds for me, till I want them. I have been promised a number of skulls more next autumn which I intend for the Smithsonian, and I hope I shall get them in safety and *sub rosa.*

I am preparing an article on the *Anatrace* found in this District for the Academy of Sciences, as I do not wish to be a silent member. I enclose it to you, and would feel *very much obliged* if you would look over it first and then forward it with the accompanying letter to Philadelphia. I shall not be able to go farther than the Geese this time, but if it be approved and published I will complete it. You will see in it my views concerning the *Berniclas.* I dare say you will find them rather startling, but they are the result of deep reflection. I began writing notes on the collections for you but I had not time to finish them.

Kennicott paid me a visit in winter, so you see he is quite a pedestrian. He was here only about a week. He is well and looks forward with much ardor and pleasure to his Summer campain in Huskey Land. He was rather in a puzzle whether to go home or stay, but science carried the day.

It may happen that I shall visit you next fall, at least I'll know when I get to Portage La Loche. 15 years are I think quite sufficient to pass in here at one spell, and 19 years out of 34 a goodly part of a mans life to have lived in the Indian territory. I only wish that I had taken up some scientific pursuit long ago. You can scarcely fancy the solace that it is to me. I am more fond that ever of it, ornithology is my great favorite.

There are much fewer small birds here this spring than usual, and I have not yet seen any new kinds. The *Plectrophanes Pictus* has not made its appearance, but there are a considerable number of *Lapponicus—nivalis* was rare. The *Dendroica Coronata* are rather numerous but not one *Spizella monticola* has shown itself. The Geese were numerous and have got a lot of necks skinned. I could not spare the whole bird. In one of these remarks the black stripe under the thorax cutting off the cheek patches one from the other as in *Leucopareias* of Brandt. The Black before skinning was *quite continuous* and broad.

I have at length begun upon a series of notices of the Dinneh Indians. Under this designation, which is a far more genuine and opposite title than either that of Athabascans or Chipewyans, I propose to include the Athabascan races of Mr. Gallatin, with some others which I consider to belong to the same family. According to my division the family consists of 5 great sections containing 41 tribes and about 32,000 souls. In this undertaking I have advanced about 8 sheets of foolscap, and am I may say only upon the threshold of the subject. Ethnology is a dry subject and I return from it with renewed zeal to my natural History Studies. I must defer writing at greater length until I receive your letter at Portage La Loche.

And with kindest regards to Mrs. and Miss Baird in which case I pose cordially yours believe me most truly yours

B.R. Ross

P.S. I have been stopped by the Sioux troubles and scarcely hope to get down until next year. Send a series of the articles to the N.H. Society of Montreal and keep one of the Birds and Eggs for myself. None of the Smithsonian publications sent me in 1861 have come to hand. They were not in the case where they were said to be and the case was full. They must have been overlooked. The *[illeg.]* american works were not be be found, nor the copy of McKenzies *travels* sent by Mr Gibbs. *[Faded]* will attempt to by it next chance.

35. RODERICK MACFARLANE to SPENCER BAIRD

Fort Good Hope 28 July 1862
[SIA RU 7215, Box 14]
[MacFarlane's nomenclature is punctuated by numerous question marks in this letter. These signify his uncertainty over the accuracy of his identifications.]

My dear Sir,

As I shall probably be pressed for time while at Fort Simpson in the Autumn, I had better now give you a brief account of this Season's operations in the egg collecting line. I may however premise that the result, from various causes, is not as satisfactory as it might have been, still enough has been done, together with the *experience* acquired by all parties, to justify me in hoping "great" things from next year's campaign among the Birds and Eggs of the "Anderson."

I left Fort Good Hope a day or two after I had the pleasure of addressing you thence, and arrived at Fort Anderson on 29 May. Our route at first was overland, and altho' Birds were seen in great numbers, no eggs were obtained until we began the descent of Lockhart River by canoe. Along its banks we procured eggs of *Falco Nigricepes?, Corvus Americanus?, Haliactus Leucocephalus and Rissa _____?* I have not been able to ascertain if the Bald Eagle nests farther North than 67 1/2° (whence we got our specimen) I should however, from information, suppose Lat: 68° to be about its extreme Northern range.

On my arrival at Fort Anderson, I at once set such Indians and Esquimaux as were about the place, to collect, and have since neglected no opportunity of directing their attention to this matter. In doing

Roderick Ross MacFarlane (1833-1920) posing for photo in 1870. Hudson's Bay Company Archives, Provincial Archives of Manitoba.

so, I have but endeavored to make some slight return for the very handsome set of books sent to me last year, and I also felt bound to do something as Mr. Kennicott was unable to come there himself, and especially as *he* had supplied me with a liberal stock of preservatives etc. for the "campaign". The accompanying list will inform you of the degree of success which has attended us; but I am far from satisfied with the result. If Mr. Kennicott had been enabled to come there, and gone down the "Anderson," and there worked with the Esquimaux, whom he would have found intelligent and industrious auxiliaries, leaving the Fort, Indian hunters, and the Eastern Barren Grounds to myself, as was arranged between us, I have no hesitation in saying that Fort Anderson would have turned out the most valuable and interesting collection yet obtained from any post in the North.

In the vicinity of the post were procured specimens with eggs of *Archibuteo Lagopus, Colaptes Auratus, Turdus Aliciae, T. Migratorius, Dendroica Striata, D. Aestiva, Ageoisthus Linaria, Spizella Monticola, Zonotrichia Gambellii?, Passerella Iliaca, Scolecophagus Ferrugineus, Corvus Carnivorus, Lagopus Albus, Gambetta Flavipes?, Tringoides Macularius?, Phalaropus Hyperboreus, Bernicla Canadensis, Dafila Acuta, Harelda glacialis, Melanetta Velvetina, Chroicocephalus Philadelphia, Sterna macroura, Colynbus Arcticus,* together with several species of Ducks, *Gulls* and other birds as per List.

I think I mentioned in my last that I intended making our overland trip E'ward to the Coast this Summer, for the purpose of collecting eggs, and if possible, a specimen or two of the Barren Ground Bear. Having to wait the arrival of Indians from Fort Simpson, I was unable to start till 19 June. We were absent 17 days, and considering the distance and the lateness of the Season, I think we did very well; we didn't however, to my regret, get a Bear. We saw fresh traces of some in the Barren Grounds, as well as along the Coast, but not an animal was seen. My party consisted of 5 Indians. We reached Franklin Bay our 9th day, spent 3 1/2 there, and returned in 4 1/2, the distance travelled in going and returning being over 300 miles. Our course was easterly thro' the wooded country to the Barren Grounds, when we proceeded N.E. Except in the vicinity of the river, there is no timber whatever to the Nor'ward of 69°; but to the E'ward of the Fort, wood of a small size is tolerably abundant, for some 40 or 50 miles, after which we entered the "Barrens", which are entirely destitute of that article. To my surprise however, I found some goodly sized timber along the immediate banks of the Horton River, in the heart of the Barren Grounds; thence to the Coast however, a distance of 40 miles, we had to dispense with fire. On the shores of the Arctic Sea driftwood was abundant. In the Wooded Country, we obtained specimens (with eggs) of several Ducks, *A. Linaria, C. Arcticus, C. Philadelphia, G. Flaupes, T. Aliciae, Dend. Coronata, L. Albus, S. Monticola, S. Ferrugineus,*

Zon. Gambellii?, P. Hyperboreus, Surnia Ulula, Gallin: Wilsonii?, Tringa Wilsonii, Plectrophanes Pictus, etc. etc.

After the first day we began to find nests of *P. Pictus,* it doesn't live in flocks during the breeding season; the male is generally to be found flying from bush to bush in the vicinity of the nest, and the female gets off the nest only when closely approached; they build on the ground in surfy commons; the nest composed of fine hay and containing from 4 to 5 eggs. We found many nests with young, on our way out, not only of *P. Pictus*, but also of *S. Monticola, Z. Gambellii, A. Linaria, T. Aliciae, G. Flavipes, L. Albus,* and *Picoides Arcticus,* etc. A start of 7 or 10 days earlier would have yielded very handsome returns. I may mention that the last two Springs set in unusually early on the Anderson; in 1857, when I first descended that river, snow was still covering the ground about the middle of June, while this year at Fort Anderson, no snow whatever was to be seen after the 20th of May.

We didn't see or procure a single specimen of *Plectrophanes Lapponicus,* until we entered the Barren Grounds, thence however to and along the Coast, they were pretty numerous, most their nests contained young, and owing also to the wary instincts of the bird, which made it get off its nest long before our approach, we succeeded in securing but a few specimens; the eggs and nest both resemble those of *P. Pictus.* The latter however has a lining of feathers which is wanting in the former. I have no idea where *P. Nivalis* builds; we didn't notice a single bird of this species on our journey, altho' for a period in the Spring and Autumn of each Season, it is very numerous along the Anderson. *P. Pictus* keeps to the wooded country, being replaced in the Barrens by *P. Lapponicus.* In the latter section, *Charadrius Virginicus* and *Numenius Borealis* were everywhere seen; but notwithstanding our best endeavors, we couldn't secure more than 3 nests of each. Very few Waders were noticed, *Tringa Bonapartii?* and *Macrorphampus Griseus*, figure in the list, as do also *Cygnus Buccinator, Anser Gambellii, B. Canadensis, Arch. Lagopus, A. Ferrugineus, Stercorarius Parasitticus?, Passerculus Anthinus?, P. Savanna?,* and *Lagopus Rupestris?.* This Ptarmigan resembles *L. Albus*, but it a much smaller bird, the male being distinguished by having a black stripe from the bill running thro' the eye; it may be the *L. Americanus* of Audubon. I have not yet compared my specimens in the description in the *Bird Book,* when it comes to hand next month, I will go very carefully over my whole collection; in the meantime Mr. Lockhart has very kindly assisted me in determining such specimens as required identification. The principal breeding grounds of *L. Albus* (in this quarter) is along the valley of the Anderson; in the Barrens however, it is replaced by *L. Rupestus?.* The Indians inform me that it is numerous, in winter, on the borders of the wooded country; if so, I shall endeavor to procure some specimens of it in winter plumage. We got but 1 nest containing eggs (4) the other nests contained from 8 to

10 young birds.

On *Horton River,* the banks of which are remarkably high and steep (composition sand, clay and gravel) we procured 2 eggs of what I believe to be *Aquila Canadensis.* The nest was found by one of my Indians, who saw the mother get off, but missed shooting her; he says that he had a good view of her, and that the tarsi appeared to be feathered to the toes. The eggs resemble those of *H. Leucocephalus*, but are not quite so long or pointed at the ends, one of them is spotted similar to the eggs of *Arch. Ferrugineus?,* (it contained a living embryo—24 June) while the other was perfectly white. I went to have a look at the nest, which was built on the face of a steep bank, some 60 or 70 feet high, and about 30 from the summit. It was erected against the banks and formed a large platform on top, (8 x 6 feet) in the centre of which the eggs were found in a bed composed of feathers, hay and moss, while the platform itself was strongly built with branches of pine, willow, etc. It appeared to have been used for very many years. I also had a view of the bird but at a great distance. We returned the eggs to the nest, and waited 3 hours for the mother, of whom nothing however was seen; and we had an examination of it on our way back from the Coast; but it didn't seem to have been re-visited.

On the Coast of Franklin Bay, as well as in the Barren Grounds E. of Horton River, we saw numbers of *Sterc. Parasitticus,* a specimen or two of which, with eggs, were secured, as also of *Stern Wilsonii?, Anser Gambellii, P. Lapponicus, C. Arcticus,* (We saw a pair of *C. Adamsii* but missed shooting them) *Aegialitis Semipalmaters, Larus* __________*?* (a large gull), and *L. Rupestris?*. You will observe that *Cygnus Americanus* is in the list; but I am not certain of this as we couldn't get the Parent Bird. We obtained No. 300 within 30 yards of the beach. The Sea, I may say, was entirely covered with ice, altho' that along shore was much broken up and somewhat decayed. We examined some 25 miles of the shore, searching for specimens, but with indifferent success; we were either too early or too late for the nesting of the Eiders etc., or they probably built their nests on the rocky Islets which appeared in the bay. You will, I think, be pleased to learn that I have secured 3 specimens of the *Somateria V. Nigra* viz. 2 males and a female. This Eider is by no means uncommon on that part of the Coast which we visited, our specimens were shot and of large flocks, and we saw great numbers of these and of *S. Spectabilis,* a male skin of which was also secured. We didn't get any *Mollisimoe.* No recent traces of Esquimaux were met with, altho' ancient houses, *câches* and skulls of the whale, were pretty numerous. Very few mammals were collected on the journey, a few skins of *Arvicolae, Myodoe* and *Arctomys Kennicottii?,* as also, of *Vulpes Fulvus* being the sum total in that line. We saw no Musk oxen, but lots of Reindeer; and great numbers of Seals and Walruses were seen on the ice, while 1 or 2 whales were heard spouting at a great rate in a narrow lane of water which separated the main pack from the shore ice. I could get but a few sea shells owing to the ice being pressed on the beach. The collection of plants will yield data as to the vegetn of the country.

On our return overland, we obtained nothing of particular consequence. Nearly all the nests we found had young ones in them, many of which were able to fly about. We however got the female and 4 eggs of *A. Ferrugineus:?* (before mentioned). The nest was found on a ledge of shaly mud along a creek we encountered in the Barren Grounds; it contained a young bird also for whose sake the male parent was spared. We got a *Myodes* No. 330 in its nest. In the Wooded tract, we procured 7 eggs of *Mel. Velvetina* (this scoter is very numerous, but it is not easy to find its nest. The female didn't get off the nest until we were within a yard of it. (exactly similar to nest of *Dafila Acuta,* in a clump of small spruce about 100 yards from a small lake.) A nest of *Falco Islandicus?* was found in a tree on our way back. It contained 4 young ones, which had only been lately hatched (3 July), 6 eggs of *Pelionetta Perspicillata* were likewise secured. On Horton River we noticed some *Scol. Ferrugineus, Turdus Migratorius, A. Linaria* and even some *Spizella Monticola.* This last builds as frequently on the ground as on a tree; it is, I think, the most common bird on the "Anderson". I fear that I have taken too many nests of it. A good many unidentified nests and eggs which resemble those of *Monticola,* are entered under this head with a ?, but may belong to an allied species. I am in hopes of getting a good skull of the Barren Ground Bear, as well as of the Musk Ox, ere the Season comes to a close.

The Esquimaux will probably bring in some good things' this autumn. Some of them brought in a lot of duck eggs without parents, which will, when identified, prove very interesting. I expected the Coast fellows will procure specimens which cannot be got elsewhere. The greatest difficulty experienced in making an Indian inland collection of eggs, arises from their having to *carry* everything, in this way, great numbers of good eggs have been lost, among others specimens of *Colys. Torquatus, Mergus Americanus, F. Islandicus?*, with some waders, etc. etc. These Indians (the Hare and Batârd Loucheux tribes) are far from being an intelligent or ingenious race; in this respect they are very much inferior to the Esquimaux, but it will go hard with me indeed, if, with the experience acquired this Season, even *they* don't do something next year.

Altho' Foxes—Cross, Silver and Red—appeared to be plentiful in the "Barrens", as well as along the Coast, we met with none of the White or Blue species, however, I think I may venture to promise a good specimen of each this winter, which will go out with the other things by the Boats of 1863. I had some thoughts of sending some eggs out by Post, but fearing that they would be overlooked, as were those of Kennicott and Lockhart last year, I have decided on sending the whole out together. We saw no Grebes, Auks, Mergansers, nor

White Wavies, on our journey.

I had better conclude now by referring you to the enclosed list, for further information, and with many thanks for past favors. Believe me, Very Respectfully Your's,

R. MACFARLANE

36. ROBERT KENNICOTT to SPENCER BAIRD

The Grove Cook Co Ill Oct 29th 1862

[SIA, RU 7215, Box 13]

[Although this letter was written from Kennicott's home in Illinois, it has been included here because of the wealth of information it contains on the Smithsonian's field program in the north. *Et li laundy saundy* translates into "which way perhaps" or was, as Kennicott stated, the Kutcha-Kutchin equivalent of the Spanish phrase *Quien sabe?* (who knows?). For translation see Kennicott's "Rubbaboo Journal" in James, *The First Scientific Exploration,* p. 119. Mary Baird's youngest brother, Charles Churchill, was, like his father, a military man. Dr. Sewall Sylvester Cutting was a Professor at the University of Rochester and Mrs. Baird's cousin. He was also a close friend of Hiram Sibley, the President of the Western Union Telegraph Company.]

Dear Prof Baird,

Yours of Oct 22 and 28th came yesterday. My Mother and father seem pleased with the idea of the Medical cadetship. They think my going into it would result in my finally studying medicine in earnest and becoming a practicing physician; thus securing a means of earning living. *Et li laundy saundy?* that is *Quen sabe?*

At any rate I shall decide upon nothing till I see you. And I shall take no steps towards entering the army till I have done some necessary *work* on the northern collections.

My Brother has loaned me some money to buy clothes with and I'll be able to go to Washington about the 15th of Nov or the 10th if you want me.

I'm *fearfully* lazy here and I fear will do little or no work till I get to Washington, where the atmosphere of industry which always surrounds you, will, I trust, have a highly beneficial effect upon me.

If the Medical cadet's duties will give time enough for work at the Smithsonian and on my report (or whatever 'tis to be) it would be a good thing. Yet if I could have the means to pay my board it might be much better to give *all* my time to my writing and Smithsonian work. But as I understand you think it right for me to use the money in hand I can safely go on as soon as you wish as it will keep me some time. It is I think $64 besides the premium I'll get on the Bill of exchange.

If I write a personal narritive besides the scientific notes the former would doubtless sell for something and so I could really get good pay for my time.

I had the pleasure of seeing Capt Churchill at Mr. Cuttings in Chicago and he was kind enough to advise me in the matter of entering the army. He told me to wait till next January anyway. I shall as I've said leave all this an open question till I see you. If I take the place of medical cadet it will be with the view of studying medicine in earnest and this could only be in the event that I gave up the idea of going into the army. But time enough to discuss this when I see you.

I left Mr. and Mrs. Ross in Red River. It was unsafe for them to come on. They will probably be here at The Grove for a day some time in November, then they will go to Canada and he will run down to Washington. That was his plan at least. He sends on a good many specimens this year and sent a good many to Europe by York Factory. I did *do* him on a fine lot of the big mountain grouse from Hardisty. That is I packed them up before he had time to select any for Europe. Hardisty of course meant them for you, so I was all right. I fear Lockharts and my Youkon specimens will be in a bad condition owing to the long land transportation last winter. The eggs are all right tho'.

Old Donald Gunns Lake Winnipeg collection is not so good as he had expected. I could have told him he was taking the wrong hunting ground. He didn't use all his money. Is going to try a better locality next spring. I can tell you all about it next spring. I've several new collectors about the Lake Manitoba region and a good prospect on the Saskatchewan.

Apropos. Chief Factor Christies in charge of Saskatchewan District said he would collect and get others to do so in his district. He said he would like a particular book; it is Gov Stevens *Report.* He wishes it on account of its treating of his own Region. Vol XII part 1st of *Pacific RR Reports* is the thing. Better secure one for him if you can. And might drop him a line Directed "Wm Christie Esq Saskatchewan District" saying you have it for him. I would like the offer of it to come from you as I think I offended Mr. Christie at Portage La Loche. Though he said it was all right. He said he would be glad to have some one from the S.I. visit him etc. and I got on swimmingly with him till I ran foul of him while defending the rights of the Mackenzie River officers. I hurt his vanity but probably he didn't care seriously (I raised myself in the Governor's estimation and did a good thing for our arctic friends, whatever).

We are *very much all right* with the new Governor (Gov Dallas).

I walked up from Peels Riv to Good Hope in Feb, bringing a big lot of our Youkon specimens. At Good Hope I found Onion and McFarlane just starting for Ft Simpson. I came up with them, stopped a few days at Ft Simpson, walked back to Good Hope, remained there six weeks, and on the 1st of June started with the boats for Portage La Loche. Thus I lost even the sight of the Anderson. I took much pains to talk with McFarlane and to teach some of his Indians at Ft Good Hope. I think he will collect largely.

Onion will collect at Ft Simpson. I took out the books directed to John Hope and gave them to Onion. John Hope would not have cared for them and would have used them about as much as he would a Greek Dictionary. We can give him other things that will please him better. I asked some of them to give him some of the novels and religious books you sent them. We have a new post office here called *The Grove.* This is now our address.

I have many things to write you about, so many in fact that I can't pick out those I ought to say first. I can't get myself into the old way of feeling yet, feel only half wakened up from a long sleep. I trust though that I'll not turn out much more lazy and stupid than I used to be.

Kindest regards always to Mrs Baird and Lucy. I hope Capt Churchill will be here at The Grove in a day or two. Yours always

ROBERT KENNICOTT

37. JAMES LOCKHART to SPENCER BAIRD

Fort Resolution, Great Slave Lake 7 December 1862
[SIA, HBC Corr Coll, Folder 26]

My Dear Professor Baird

Your very kind letters of 31 March, 4 and 14 April, have all reached me in due course. Your "small shot" and "heavy artillery" have successively hit the mark. The Books are slowly working their way into my head, the "small shot" penetrated my waistcoat pocket, but the *whiskey,* that went straight to the *heart*! I know not well how to express in appropriate terms my sense of your kindness, but like the Negro's Parrot, altho I can't speak much, I "tinks a great deal." Nevertheless accept my warmest thanks for the kind attention you have shewn to all my wants and wishes. The Maynard Rifle is a perfect gem; out of all comparison the most commodious and efficient arm that I have ever seen, or even heard of. Mr. Kennicott insists that I should accept it, as a parting gift, a "souvenir," from himself, and I have not the heart to pain him by a refusal.

I now forward, for your information, a List of what I have collected, since last I wrote to you. I am sorry that there is so little in it, that will be new to you. I did my best, however, and I flatter myself, there are a few nests of Eggs, which you will be pleased to see. I think I have succeeded in obtaining several of those specified in your last list of "Northern Eggs yet to be collected," which Mr. Kennicott forwarded from Portage La Loche. Mr. Macfarlane has also collected several new Species at the Anderson among which I noticed *Plec. pictus* and a number of Terns, Gulls, and ducks which I was unable correctly to identify. The Anderson is now the place from which you will receive the most valuable specimens and Eggs. It is a thousand pities that Mr. Kennicott could not carry out his original intention of passing a summer there. He has, however an admirable substitute in Mr. Macfarlane, who has now caught the Oölogical fever, and will, when business permits, spare himself no trouble in working up that field. I am sure you will be delighted with his collection of last summer. He told me that he would write to you himself, regarding his collection by this packet.

I am very sorry that I could not remain another year at the Youcon, for the benefit of Science. That place is not yet properly worked up. We have received, as yet, next to nothing from any of the distant tribes. I sent word, last winter, to some of them to collect Specimens and Eggs for me, but neither the Gens des Bûtes, from down the Youcon, nor the Gens de Fous, from above had arrived when I left for Fort Simpson. I arranged with Mr. Jones to take everything they might bring. Mr. Jones also promised to "go in" to the Fishes of the Youcon Valley. I have left with him, for that purpose, all the Alcohol, alum and Saltpetre, the Seine etc. etc. which Mr. Kennicott sent to me, and I am sure he will do whatever he can.

Great Slave Lake has been so much worked up for the last few years, by Mr. Clarke at Fort Rae, Mr. Reid at Big Island and Messr. Hardisty, Kennicott and McKenzie, at this place, that I fear there is very little left for me to do. I am not without some faint hopes however of being able to add a little to that already collected by my predecessors.

Your enquiries respecting the Goats or Sheep at Peels River and Lapièrre's House are easily answered. During my residence of two years at Peel's River, I made many voyages across the mountains to Lapièrre's House, and have frequently met in with herds of these animals, but I never saw any with "black" horns. I saw many with large *white* horns much curved or twisted, and was told by the Indians that these were the males, whereas the females have the horns such as you describe them, i.e. "some six or eight inches long slightly curved backwards." In every Loucheux lodge there are several spoons made from the horns of these animals, but I never saw a "black" one. The Gens des Fous from up the Youcon, on their appearance at the Fort, generally have each a horn spoon in his canoe, to use as a drinking vessel; some bring spoons and entire horns for sale to us, but I never saw a "black" one. The inference to be drawn from these facts therefore is, that there are none of the "black" horned species, or that they are exceedingly rare, in the Mountains at Peels River and Lapièrre's House, or in the Gens des Fous mountains, which are a continuation of the same range and distant ten or twelve days march. I have learned, however, that the black horned species is numerous at Fort Halkett and have therefore written to Mr. Brass to endeavor to procure a few specimens.

Mr. Kennicott will have reached home doubtless, long ere this, and will have given you full information regarding our doings and prospects in the Natural History way. I need not therefore tresspass further on your valuable time, by a repetition which I always disliked.

Many thanks to you for your kind offers to be of use to me.

I remain, with much respect, Very Sincerely Yours,

J. LOCKHART

38. RODERICK MACFARLANE to SPENCER BAIRD

Fort Anderson 6 May 1863

[SIA, RU 7215, Box 14]

[Lewis Henry Morgan (1818-81), the noted American anthropologist who studied Iroquoian culture and wrote *Systems of Consanguinity and Affinity of the Human Family* (1871) and *Ancient Society* (1877), used the Smithsonian distribution network to disseminate a "Circular in Reference to the degrees of Relationship among different nations" during the 1860s. He used this questionnaire to collect information on indigenous kinship systems.]

My dear Sir,

I have to acknowledge receipt of your esteemed favor of 23 Octr. last, which reached me on the 24 ulto, and in reply beg to express my thanks to you for the kind manner in which you speak of the comparatively few objects of Natural History I had up to that time collected for, and forwarded to, the Smithsonian Institution. I shall however try and do what I can this season, as I may not have another opportunity afforded me for doing so, in this quarter.

We were all disappointed at not receiving letters from Mr. Kennicott; but at the same time extremely happy to hear, thro' you, of his safe arrival at the Grove.

Since I had last this pleasure in February, I have packed up and forwarded by sled to Fort Good Hope, 7 cases addressed to the "Smithsonian Institution," while the eighth (No. 1 of Packing Account) which I have not yet filled up, is retained here until I start the Spring Packet ten days hence, in order that any *rarities* of this season's collection, I may, in the interval, succeed in procuring, may then be forwarded, and thus reach you at the same time with last year's collection. I have already obtained 12 eggs of the Hawk Owl *(Surnia Ulula)* (2 nests containing 6 eggs each!) also another nest which contained 6 eggs—but 1 broke—and is supposed to belong to the same species; likewise the female bird, nest and 4 eggs of the Whiskey John (*Perisoreus Canadensis*); but the eggs of this specimen, as well as a great number of last year's collection, are in a disreputable condition, owing to the large embryos they contained. I trust however that I'll be enabled to replace some, if not all of them, with better specimens, ere the Season comes to a close.

With regard to the contents of the above mentioned cases, I fear that they will not afford you so much satisfaction as my previous lists and letters would have led you to expect, while the list with relative notes, now forwarded will not be found to agree exactly with these. In explanation, I may mention that some specimens were subsequently found to be too much damaged to forward, while others were not forthcoming, and some even appeared to have had their labels accidentally changed; it is also probable that mistakes were made in course of the packing, as I was at the time, myself, far from well, and therefore unable to give that attention to the business which, to make objects collected really valuable to Natural History, I am well aware, is absolutely necessary. My previous inexperience will also, I trust, plead for my shortcomings, especially when I assure you that I shall endeavor to make my collection of this Season, as deserving of your *perfect confidence,* as I can possibly make it. At the same time, I may mention that this is not so easy as might at first sight appear, owing to the stupidity of the Indians etc., who can rarely be trusted to do anything correctly, when acting by themselves and beyond one's personal supervision. Whatever may be collected by myself, or under my immediate direction will of course be pretty sure of being as well identified and authenticated as possible. Some specimens are also, unavoidably, retained for transmission next year, while others have been sent out which are likely enough to prove of little or no interest to Naturalists.

I subjoin a list of documents which will be found in Cassette No. 1, the key of which is herewith transmitted. These documents comprise Registers of the Weather, lists of Esquimaux Curiosities forwarded, together with a copy of my "Note Book" to the present time, all of which I trust will reach you in safety, and prove satisfactory.

I regret to say that I have been unable to procure a mountable specimen of the Blue Fox; but as I have now spoken to *all* my Esquimaux on the subject, I think you may safely rely on getting several good specimens next year, together with as many more of the *Vul. Lagopus* as may still be required. I now however forward a *fur* skin of a female Blue Fox which was killed on the Arctic Coast this winter. The Esquimaux recognise it as a distinct species, which the skin itself either in Summer or Winter dress will, I think, go far to prove; they call this Fox *Ighaluk,* and *Lagopus Terrayana.*

In order that you may obtain a correct idea of the different kinds of furbearing animals which abound in this quarter, I beg to enclose a list of the Fort Anderson Returns for the current Outfit, premising however that this is but our *second* year, and among a people who are now only beginning to hunt with the aid of Traps and Guns. For the sake of comparison, I append an abstract of our first year's Returns.

From a party of Esquimaux who have just arrived, I learn that one of my chiefs has killed a large Barren Ground Bear which they say he has skinned according to the directions I had previously given him; if this be true, I dare say the specimen will be better adapted for mounting than any yet obtained. Another specimen of this Bear (No. 540) lately procured from an Indian was unfit for this purpose; but the skeleton, being entire (*minus* the shoulder blades) and in good order, shall be transmitted by the Boats of next Season.

I shall start (*D.V.*) on my proposed trip to Cape Parry as soon as ever I can arrange my Esquimaux, when they come in by open water, in the spring. I should like to be off by the *10 June*, and sincerely trust that the expedition will prove successful, even beyond my expectations, and they are sufficiently high. I may also mention that I have just received from an Esquimaux, five indifferent female skins of the Eider (1 of the *Som. Spectabilis?* and the other is *S.V. Nigra*, or *Mollisimae?*), together with 12 eggs, but as these were not blown at the time, in fact only now by myself, and are consequently much cracked and discolored, they can hardly be considered as good specimens. They were found last summer on an Island in Liverpool Bay, and the parent birds were said to have been shot in the vicinity of the nests. It will go hard with me indeed, if I fail in getting good specimens of all the Eiders, together with their eggs, in course of the ensuing campaign.

With regard to Mr. Morgan's Circular, I may inform you that I have reason to believe that the peculiar system of Relationship discovered by him to exist among the Indians of the United States, prevails also in this District, at least, I have ascertained, to a certainty, that it obtains among the Indians of Liard and Good Hope; but for want of a good Interpreter, I cannot as yet say that it is likewise common to the Loucheux and Esquimaux. I shall however endeavor to fill up schedules for each of the above mentioned nations, in time to forward from Fort Simpson in the Autumn; together with the printed vocabulary of words (filled up) received last year.

15 May '63

Since writing the preceding, I have received 6 more eggs of the Hawk Owl, some of them however are considerably cracked; also 4 eggs with the parent of *Corvus Americanus,* with 6 eggs of *C. Carnivorous*; this is merely supposed in consequence of the Indian who brought them having asserted that they belonged to the larger crow. Nest found in a tree, while the eggs resemble those of *Americanus.* I have also obtained 4 eggs of, I think, the *Falco Islandicus.* i.e., judging from the description of the bird given by the Indian who found the nest, and missed hitting the parent; he says she was a bird of a goodly size, of a light color, the tail long and barred with brown, and that this species generally prey on Partridges. Nest was found on a tall pine, composed of sticks and twigs, and lined inside with dried mosses; grasses, and a sprinkling of feathers. Eggs are of a reddish color, similar to those of *F. Sparverius*, but considerably larger than even those of *F. Nigriceps* of the List.

The Packet is just to start, I must therefore conclude with a hope that when I again have this pleasure, (*D.V.* in August) I shall have important results to communicate.

I remain, Very Respectfy Your's

R. MACFARLANE

39. WILLIAM MACTAVISH to SPENCER BAIRD

Fort Garry 21st May 1863

[SIA, HBC Corr Coll, Folder 33]

[The Reverend Griffith Owen Corbett was a member of the Colonial and Continental Church Society. He went to Headingly in 1854 when a parish was established there, and he stayed until 1863 when he was forced to leave. In this letter Mactavish expresses his disgust over the events which precipitated Corbett's departure from the district. Corbett was charged, tried, found guilty, and imprisoned for seducing and then attempting to procure an abortion for Maria Thomas, a domestic servant living in his household. The ruling of the civil court was later reiterated by local ecclesiastical authorities, but Corbett and his supporters protested both charges. These developments had, as Mactavish points out, repercussions for the Institute of Rupert's Land. The leading executive members of the Institute were all clergymen—the President was Bishop David Anderson, and Archdeacon Cockran, Archdeacon Hunter, and Rev. John Black were Vice-Presidents of the scientific society. Andrew Graham Ballenden Bannatyne (1829-89) left the Orkney Islands when just fourteen, to become an apprentice clerk with the HBC. He subsequently became an important merchant and community leader in Red River. He established a dry goods store soon after leaving the company in 1851, and became a strong supporter of Metis rights, even assuming the position of postmaster in the provisional government of 1869-70. Hector Mackenzie (1816-89) was born in Rupert's Land. He signed on with the HBC as an apprentice Post Master in 1839 and retired to Red River in 1851. He spent most of his career in the Athabasca and Mackenzie River districts. Roderick McKenzie (b.1824) of Lochbroom, Rosshire, joined the HBC in 1847 as an apprentice Post Master. He was stationed at Manitoba from 1856 to 1861-62, when he was transferred to Cumberland House. He was promoted to Chief Trader when he relocated.]

My Dear Professor Baird,

I have so long been silent that I am almost ashamed to write you, but I have *as* you know a good excuse, as I have all kinds of trouble and annoyance in addition to my constant worry that is full occupation. I feel too much disgusted with the politics of the place to enter into them and I daresay you are no way inclined to be bored on the subject. I have whole hosts of your letters lying before me which appear to have not been answered, but I need more especially only refer to your two last, both of 11 April, which came to hand some time ago with a great many of your letters for your northern correspondents, they will go on in a few days.

Clare and Christie were both at home the winter before this last so their collections could only be made this winter. I have not heard what either of them have done, but I got a can of specimens from York last fall. At least I was told there were specimens of all kinds in it, for to tell the truth, I have not opened it yet. This, with whatever else may reach me, I must get sent to you by some chance to St. Paul.

I suspect Mr. McFarlane has been misinformed when he says the Governor issued a general order for all specimens to be sent to England. Any way I will speak to Mr. Dallas on the subject of the specimen of the Barren ground Bear for the Smithsonian and have no doubt, if it can be found out, to have it taken from the specimens for England and sent here for you from Norway Ho.

It is but too true the Rupert's Land Institute has blown up like everything else in the Settlement. The Church not excepted. The Parsons have split among themselves and the people here have taken sides according to their feelings. Of course a most unreasonable mess is the result. The origin of the quarrel was a criminal accusation against a clergyman who was popular in his parish. How he managed to carry his people with him I cannot say as he was a most inferior animal, but the people here are desperately ignorant and it seems to me he pandered to them. He was tried, found guilty, but his partizans would not give him up, but broke up the Gaol and took him out. Though unfrocked by the Bishop he still continues to preach and his people are determined to protect him against the authorities, who to avoid bloodshed have for the present allowed the matter to rest. Of course the members of the Institute could not meet, but before this took place our only newspaper has shown itself hostile to the Institute as the Editors could not lead being grossly deficient in science. I have no hope that it will be set going again in a hurry, though I daresay some of the members will continue to collect. I notice your wants and will endeavor to meet them. York is a capital place for Grey geese and I will send there for a series of specimens. There is there a very large grey goose commonly called "barren geese" which come to the coast after the others have passed Northwards and there is another not larger than a stock duck. I have never seen the Blue Wavy at York though I have killed it about 100 miles up the river. Don Gunn purposes going out towards Manitobah this spring if his wife's health improves. She is at present very low and the old Gentleman is in very bad spirits about her. If Kennicott goes to the Hudson's Bay Coast he should look closely along the coast between Owl River and Churchill. About Cape Tatnum to the South of York is also famous as a Breeding place for gulls as there are a number of Islands there called the "Pens." I have a lot of Beetles which I will send you from here and along Lake Winipeg, besides a few other Insects and a large bottle full of Insects from Fort Simpson. I had intended to examine them this winter but other matters and sickness have prevented me. I shall be glad when you publish your work on the Natural History of this part of the world, and I hope if any more of them are published you will not forget to send me the continuations of the works you sent me last spring, particularly on the *coleoptera.* I daresay I may get Mr. Bannatyne to prepare some specimens for you but he is a dilettante collector and generally wishes to set up specimens to astonish. How lucky when they come into his rooms. Hector McKenzie is more of an out and out Collector, and restricts his attention almost exclusively to Birds. By the way, in the list of specimens taken down by Mr. Kennicott from me, I see he has confounded the Collections sent me by Roderick McKenzie from Manitobah, with those of Hector McKenzie of this place. I may as well hint to you that Governor Dallas wishes to have specimens for a museum he intends forming at his place in Scotland, or rather which he intends having in Scotland, so that you will have time to consult him as to his requirements. I mention this as you ask me what you can do for him. In haste, Believe me Yours faithfully

W. MACTAVISH

[Postscript]

The accounts from here against the Smithsonian for supplies to Mr. Kennicott was sent you some time ago, and I believe the accountant here wrote you with it. I would write you soon, so I have intended, but I was for about six weeks confined to bed and work accumulated in the meantime. However, the full account was sent and details of all the changes except one which will also be sent as soon as I can get it. Should there be any error you can point it out with Mr. Kennicott's assistance. He spoke to me about some overcharge in freight but I could not say anything about it as Sir George Simpson has not communicated to me the instructions on which Mr. Kennicott claimed the reduction. However he can state his objections and the charge will be settled, I am sure, in a liberal manner. The Rev. W. Kirkby wrote me that he owes the Smithsonian for printing but sent me a list to remit to you. But it was so perfectly ridiculous in amount that I returned it to him and pointed out that it was so deficient in amount that he must have made some error.

W. MACTAVISH

40. RODERICK MACFARLANE to SPENCER BAIRD

Fort Anderson 13 July 1863
[SIA, RU 7215, Box 14]
[Ed. Harriott was MacFarlane's apprentice clerk during the 1862-63 outfit. (H.B.C.A., PAM, B.239/k/13)]

My Dear Sir,

I returned here y'day from my journey to the Arctic Coast, and as I start in a day or two for Fort Simpson, where I shall probably be pressed for time, I shall now address you on the result of this year's collections in Natural History.

On the return here, on the 18 ulto, of two of the men I had sent in to Good Hope last April with some of the cases containing last Season's Collections, I learned, with much regret, that 2 of these (Nos. 3 and 4) had been left along the road, owing, as they asserted to the thaw having set in so very early that the dogs could not get on at all. This might have been so; but I know that more difficult things are frequently performed in the North. Mr. Gaudet also wrote to me that he could not send for them, had he however been enabled to send back here the men at once as I desired, I should thus have been made acquainted with the circumstance, and the cases would now, instead of remaining behind till next year, be on their way to Washington. I shall however endeavor to render them at Good Hope now, and for the future, I shall certainly guard against the recurrence of so provoking a thing.

The Spring, as mentioned in a previous letter, set in unusually early; but the weather became cold and inclement afterwards, and with few exceptional days has so continued since. This prevented the Esquimaux from resorting to the Fort until the middle of June; in fact, I got them off only on the 20th, so that I was thereby prevented from starting on my Coast journey until the evening of that date, 10 days later than I had proposed. It was impossible for me to have left earlier, owing to the complement of men having been reduced to 6, and the retirement of Mr. Harriott who was here last year. I need not state that if I had been enabled to carry out the arrangements mentioned in my May letter, the results would have been superb. They would have been considerably larger than they are, while nearly all the eggs would have been *fresh* and therefore *good* specimens. As it is, a large proportion of those collected are very indifferent specimens; but I sincerely trust that another season will prove propitious in all respects.

The enclosed list will of itself furnish all the information you may wish to know until the collection goes down; but I may also as well make a few remarks in elucidation thereof.

The SUPPOSED *additions* in Oölogy comprised in this season's Collection, embrace specimens of *1. Buteo Borealis, 2. Archibuteo Sanctijohannis, 3. Brachyotus Cassinii, 4. Picoides Arcticus, 5. Cotyle Riparia, 6. Collyrio Borealis, 7. Macrorhampus Scolopaceus, 8. Calidris Arenaria, 9. Micropalama Himantropus, 10. Tryngites Rufescens, 11. Numenius Hudsonicus, 12. Somateria Mollissima, 13. S.V. Nigra, 14. S. Spectabilis, 15. Ersimatura Rubida, 16. Mergus Serrator, 17. Stercorarius Cepphus, 18. Larus Glaucus, 19. L. Leucopterus, 20. Mareca Americana, 21. Xema Sabinei, 22. Colymbus Septentrionalis, 23. Perisoreus Canadensis, 24. Fulix Affinis, 25. Passerculus Sandwichensis.*

I may however mention that I suspect that the eggs of several of the above birds were obtained last year and sent out among the unidentified specimens, while many of them are but poorly represented. In fine, much yet remains to be done in Arctic Oölogy, especially among the Ducks, Geese, Waders, Gulls, etc., but another year may probably yield further, and I trust, more important additions to this most interesting branch of Natural History.

If the eggs of the Golden Eagle (*Aquila Canadensis*) be still the desiderata they were considered a year or two ago, you will be pleased to learn that I have secured 7 *well identified* specimens, that is, 3 nests containing 2 eggs each, and the fourth 1. Nos. 553 and 562 of list were procured—the former on 21 and the latter on 27 May, and the eggs contained well developed embryos—from the Barren Grounds to the N.N.W. of Fort Anderson, and the female parents were snared on the nests, these were found built on the face of a steep bank, and in construction similar to that described under No. 285 of list. No. 605 however was found (8 June) on the top of a tall pine, near a small river which joins the Anderson, some 25 miles to the N.E. of this. The nest was composed of a large mass of dried twigs, willows etc., lined with hay and a few feathers. The female was seen very distinctly, fired at and fell, but could not afterwards be found. The Indian hunter assured me that it was the Golden Eagle, precisely similar to the skins here, but rather darker in color. He also stated that it was not the white headed bird, and that he once before knew of *A. Canadensis* building on a tree. The country in that quarter being sandy, therefore unsuited for building, will probably account for this deviation from its customary habits. The nest appeared to have been used for several years. The eggs of these 3 nests are nearly alike, and while agreeing pretty well with Audubon's description (except in size) in Brewer's *Oölogy*, are somewhat different from those obtained last year (285) altho' these also undoubtedly belonged to *A. Canadensis* as will hereafter be shown. No. 677 was obtained from an Esquimaux, and the female was shot on the nest. The same having been built against a high bank in the Barren Grounds near the lower Anderson, in perhaps Lat. 69 1/2° N. No. 661 is the skin, merely, of a female bird of this species, which was shot on its nest by another Esquimaux—2 eggs unfortunately broke—one of which was in the oviduct of the female. No. 1014 is the male skin of a bird shot by us on the Arctic Coast. I forgot to mention that the egg of 677, is smaller and differently marked. It very much resembles that of *Halioetus Leucocephalus.* As *A. Canadensis* is by no means very scarce in the "Anderson Territory", I think I may venture to promise 2 or more specimens of its eggs next year.

I am of opinion that the 3 species of *Archibuteo* described in Vol. 9 of the P.R. *Reports,* nest in this quarter. You will perceive from the List that a good many specimens were secured, many, I should say most, of the nests were found on trees and the others on cliffs; I have entered them under *Archibuteo,* altho' it is quite possible that some of the eggs may really belong to *Buter*____?. I regret to inform you that the eggs of the large White Owl (*Nyctea Nivea*) have hitherto eluded our search, also the nests of (*Ampelis Garrulus;* but I am in hopes that the season of 1864, will make them known to us. *Perisoreus Canadensis* I trust will also be better represented and doubtless there are many others, even among the Smaller birds, nesting in this quarter, whose eggs are yet destined to figure in future lists.

It was so late before I could leave for the Coast, and the time being limited, that I, from the outset, relinquished the idea of reaching Cape Parry this season. The road I followed myself was the same as last year; but as my party was sufficiently large, I made 1 Servant and 2 Indians follow a paralell course to my own to the Nor'ward of my track, as far as Rendezvous Lake, on the borders of the Wooded Country, while another party pursued a similar route to the S'ward. On meeting, the Oölogical result was as followed viz.—myself and party 70 specimens, chiefly *Plectrophanes Pictus* (13 spec.?), *Spizella Monticola, Zonotrichia Gambellii, Dendroica Coronata, D. Straita, Turdus Aliciae, Lagopus Albus, Scolec. Ferrugineus, Phalar. Hyperboreus, Gam. Flavipes, Colymbus Arcticus,* etc. The party to the North procured 50 specimens of *P. Pictus* (16 spec.?), *S. Monticola, Z.*

Gambellii, T. Aliciae, L. Albus, L. Rupestris?, P. Hyperboreus, Micropa. Himantopus, Tringa Wilsonii, Passerculus?, Chroico. Philadelphia, C. Arcticus, Archibuteo?, etc. And the other party got but 20 of *Colymbus Arcticus, Archibuteo?, L. Albus, P. Pictus (2 nests), S. Monticola, Dafila Acuta* etc. It would thus appear that *P. Pictus* is rather a scarce bird to the Southward of the route I followed myself, while it, as well as several other species, are more numerous to the N'ward. Owing to the lateness of the Season, when we started, many of the eggs are very indifferent specimens, owing to the large embryos contained in them, while a scarcity of Alcohol prevented me from securing any number of Alcoholic specimens. We found *L. Albus* to be more numerous in the vicinity of the Fort, as well as on the journey to the Coast, than was the case last season—about 950 eggs having been obtained during the Season—800 more than in '62. In fact, white partridges were seen, and specimens with eggs, procured by us along the entire route, and some birds were also seen on the Coast, altho' not a bird was noticed in the Barren Grounds proper on last year's trip. Most of the specimens are well identified, many of the parents having been snared on the nest. I may however remark in passing that to me, the eggs if not also the skins, seem to indicate at least *two* species of *L. Albus. L. Rupestris* (was likewise met with from the Borders of the Wooded Country to the Coast, they were however most numerous at the latter place. We managed to secure a few well identified specimens of their eggs, their nests are extremely difficult to find, and in this respect, rank with *Ampelis Garrulus,* and *Perisoreus Canadensis* etc.) Regarding this Partridge or Ptarmigan, it may also be mentioned that the eggs lead me to believe in the existence of another and a fourth species, but the results of another season are however necessary, in my humble opinion, to determine the point to the satisfaction of Naturalists. A Scarcity of Ammunition and Arsenic prevented me from securing male specimens of the different kinds of *Lagopus,* as well as of other birds mentioned as desiderata in some of your letters. The egg (No. 10 of last year's list) I now believe to have belonged to *L. Albus,* not withstanding its color (entirely white) as I have this season taken 2 similarly color'd eggs from the oviduct of female birds of this species (Nos. 566 and 573). The coloring matter would appear to be deposited very shortly before the emission of the egg. I had proof of this on examining the oviduct of a female snared on a nest containing several eggs having the usual markings; it contained an egg perfect in form and marked like the others, the matter however was so soft that it adhered to the fingers. Another egg (No.) found in a nest, was extremely small and ovoid in form, and scarcely an eighth of the dimension of the others contained in the same nest. *L. Albus* lays from 7 to 13 eggs, the general number being 8, 9 and 10. *L. Rupestris* lays from 4 to 8, rarely more, and all the eggs of both species are hatched; that is, I have every reason to suppose so, altho' I believe that the young do not all arrive at maturity. Some nest early and others again very late in the Season; *fresh* eggs were found by us after we had seen young birds of the different species. One nest of *L. Albus* was found within 70 yards of the Fort!, and others at distances varying from 130 to several hundred and thousand yards.

Altho' the striped Woodpecker (*Picoides Arcticus*) is not scarce in this quarter, we have succeeded in finding but one nest of it containing 3 eggs—not the full number I should say—the nest was in a hole in a dry pine and about 4 feet from the ground. A nest of *Colaptes Auratus* with no less than 11 eggs, was also found in this vicinity.

On a small lake near a river flowing thro' the Barren Grounds, we found the nest of a Swan, (*Cygnus Americanus*), the male of which was shot and the head etc. retained for identification. The bill is marked with a spot of yellow in front of the eye; the nest was on a low flat swampy island, in the middle of a small lake, and was composed of a large raised up mass of turf and moss etc. Both parents were seen disporting themselves on the lake, and the female bird was fired at and missed.

On our arrival at Horton River, we at once proceeded to the nest of the Eagle from which 2 eggs (No. 285) were procured 23 June '62. We were all together at the time, and seeing no signs of the bird on reaching the nest, we concluded that she had abandoned it, we however sent one man up to examine it, never expecting to find it occupied after the noise made below, and were consequently unprepared for the flight of the bird which took place only when he had reached it. A couple of shots fired in the surprise of the moment took no effect, the bird having flown away without even uttering a cry; or making the slightest attempt to defend her offspring. The nest was found to contain a young eaglet of about the size of a Partridge, covered with a thick whitish down, the rudiments of the primaries and secondaries of the wings, also the tail feathers, were slightly tipped with brownish black. The nest contained the greater portion of a Partridge besides an entire marmot [Baird wrote in *Spermophilus parryi* at this point]. As my party included the Indian who had succeeded in snaring the Eagles (Nos. 553 and 562) I got him to set a couple of snares, and as it was about camping time, we rested for the night at the distance of a mile and within sight of the nest. The weather changed during our stay from being fine and warm, to rain, wind and cold. Next morning, we of course expected to find the mother secured, but to our surprise we discovered that she had not returned to the nest, we therefore supposed that the shots fired at her had taken effect, and were at the same time so disgusted with her cowardly conduct, that I ordered the snares to be left on the nest until our return from the Coast, thereby making sure of her capture should she be still in life. I was the more anxious on this point in order to *prove* the parentage of the eggs obtained last year. The nest was reached on 25 June, only 2 days later than in 1862, still the eaglet appeared to be about 2 or 3 weeks old. I have however remarked this season that all the *Raptores* and many other species accustomed to nest early in the season, did so earlier than usual this spring; this I attribute to the extreme fineness of the weather (for the season and situation) during the months of March and April. The cold and lingering weather afterwards, had the contrary effect with many of the other birds. In fact, the nesting of birds in these parts greatly depends upon the state of the weather in the spring of the year.

On our return to Horton River 9 July, after an absence of 14 days, I proceeded with one of my men to take possession, as I expected, of our prey, and were again when almost on the nest, startled by the flight of the bird, the shots therefore fired at missed her. She never uttered a sound but flew quietly away as if there had been no young one in question. On examination of the nest we found that the eaglet had greatly increased in size. It could not however stand upright, but managed to support its body on the tibiae and belly; it offered to bite us when closely approached. We also observed that the female had been caught (by the leg I presume) in one of the snares, and that she had after much trouble succeeded in liberating herself by cutting it! I got the nest measured, the dimensions being as follows, vizt., extreme length 13 feet 5 ins, breadth 10 feet, height in front 7 ft., increase since last year chiefly in height. I need not say that I was astonished at the total want of courage and parental devotion displayed by this Eagle in defence of her offspring. Of all the birds I have ever seen, this one behaved the most cowardly; in 1862, she flew off the nest when approached, and was seen hovering for a short time in the air and then disappeared, this summer she flew away as cooly as if the nest contained neither egg nor young, and did not return while we were in its neighborhood, altho' the weather was very cold, and on our return, when the nest was approached only by two persons, she again disappeared without having evinced a spark of parental affection for her young. We dined at our encampment of 25 June, but saw no more of her. I am however (*D.V.*) determined on having her as well as her eggs of next season, if she again return to build there. I think that such strange behaviour, the reverse of the eagles in Europe, would go far of itself to warrant the belief that the species are entirely distinct. The young eaglet stood the trip in very well and is increasing daily in size, eats meat and fish and partridges, with a preference however for flesh. Her plumage has undergone a great change since 25 June. At present, she is rather larger than a White Wavy, crown of head lightly tipped with brown, the back feathers black, the first whitish plumage still occupying the greater portion of the body, primaries, secondaries and tail feathers black, these however are still short. Bill bluish, cere bright yellow, toes also bright yellow, the tarsi thickly covered with soft whitish down to the toes. Another eaglet of the *Haliaetus Leucocephalus* obtained by an Indian a few days ago, is also improving daily in appearance, eats fish and meat with a good appetite. The Indian found the nest near Lockhart River, and in order to get at its contents (3 young eaglets) he was obliged to cut down the tree, whereby 2 of them got killed. The parent bird showed no great spirit in their defence. This eaglet is much larger than that of *A. Canadensis*, and is also entirely different in appearance. These once seen, no one could ever mistake one for the other. The under plumage is a soft velvety *grey*. Darker on the back—the primaries, secondaries, and other feathers light brown—Bill dark blue, cere same color—sides of bill deep orange yellow—the tarsi in front covered with grey down to the middle of same, naked behind—is a very quiet timid bird, the very reverse of the above mentioned; both however agree remarkably well together, and they are now able to stand on their legs and also walk about. I shall write you more fully anent them and other matters from winter quarters. I doubt however if I can protect them from the cold during winter. I shall try; but at present I cannot express an opinion as to their future disposal.

In the Barren Grounds to Horton River we secured several specimens with eggs of *Charadrius Virginicus, Numenius Borealis, Tryngites Rufescens, Stercorarius Cepphus,* etc. On 29 June we crossed thence to the coast, and altho' it blew a perfect gale with snow and was also very cold, still we managed to collect a few specimens of *T. Rufescens* and *Lagopus Rupestris*, also of *Calidris Arenaria*, and another of *Tringa Bonapartii*. Last year we did not notice a single bird of *T. Rufescens*, while no less than 12 specimens figure in this summer's collection.

The coast was reached a few miles further northward than last year; the Sea was more covered with ice than was the case at an earlier date last season. On 30 June we proceeded down the bay but found very few specimens, the Foxes having anticipated us in one visit; on 1 July, we encamped on the point which formed the extreme limit of our journey in '62. Numerous bands of Eiders, especially *S. V. Nigra*, from their large size, were seen, but none could be approached close enough for a shot, we however secured the skeleton of a male which had evidently been killed by some other bird. With the exception of the 30 June, the weather during our stay (9 days) on the coast was very unfavorable for operations of any kind. Wind, snow, sleet and rain with heavy fogs prevailed during that period; it was also extremely cold. In these circumstances, and owing also to our late start, the idea of being enabled to reach Cape Parry was reluctantly abandoned, still a considerable portion of the Coast in that direction was examined; but I am sorry to say with little success. Foxes etc. had been beforehand and destroyed a considerable number of eggs of Swans, Geese, Ducks, Plovers, Gulls, Terns and Partridges. On 3 July however we found the nest of an Eider duck (*S. Mollissima* I think) on the mainland, the female of which was shortly afterwards secured on the nest with a snare; it contained 5 eggs, and was a mere depression in the ground,

200 yards or thereby from the Sea, lined with down. I am not quite satisfied, I may state, that *S. Mollissima* exists on this part of the Coast. I incline to the opinion that it does not, notwithstanding the number of specimens subsequently received and entered under this head, they probably belong to *S. V. Nigra.* In my letter of last August, I mentioned that several Islands lay off the said point, which I was last season unable to visit for want of a Canoe; this induced me to take on a couple of Esquimaux with their Cayaks for the purpose of examining these Islands for eggs. As the Sea was so much covered with ice, and owing also the state of the weather, but one Island could be visited and this was done by them at no little risk. Unfortunately however we were again too late; the Foxes or some other animals having destroyed a considerable quantity of eggs, still they succeeded in getting a few specimens of *Somateria, Larus Glaucus, L. Leucopterus, Xema Sabinei, Colymbus Arcticus, C. Septentrionalis, Phalaropus Hyberboreus, Tringa Wilsonii, Anser Gambellii,* etc. This Island formed one of a group, and was several miles in extent, some Reindeer were also seen thereon. I regretted that I could not visit it myself, Esquimaux Canoes can only be made use of by the Esquimaux or others after long practice, as they are very cranky, while an upset is sure to drown one if no aid be at hand. In the open water which appeared at spots along the Arctic Shores, we noticed great numbers of Divers, (*C. Articus* chiefly) Ducks (Scoters and Eiders) and other Sea birds. Gulls and Terns were likewise numerous, and Sabine's Gull by no means scarce. We secured 4 specimens of it, one with 3 eggs, this last was found on the Island visited by the Esquimaux. On the Coast we also shot a large Barren Ground Bear, and a small seal. Some Walruses were seen but at too great a distance to shoot. As I shall again address you by Winter Packet more fully and intelligibly than I now do, I had better conclude this long and uninteresting letter, ere doing so however, I may mention that on our return overland to Fort Anderson, we procured several specimens with eggs of *S. Albus, L. Rupestris,* and other birds. From the list you will perceive that a good many specimens of *Charadius Virginicus* have been secured. This bird certainly nested much later this year, as many eggs found at a late period of the season were tolerably fresh. The nest of *Numenius Borealis* is not easy to find; on our way back we noticed a great many young ones of this species, also of *P. Pictus*, and other birds. In the Barren Grounds we also procured a few specimens of lemmings and *Arvicola.* On my arrival at F. Anderson I found that the Esquimaux and Indians had brought in a good many "fine things"; but as I shall write you fully in Winter on all subjects, I had better subscribe myself, With much esteem,

Your's Very faithfully,

R. MACFARLANE

41. JAMES LOCKHART to SPENCER BAIRD

Fort Resolution 15 December 1863
[SIA, HBC Corr Coll, Folder 26]

My Dear Professor,

I have to acknowledge receipt of your favor of 10 April, by the Portage La Loche boats this summer.

I am pleased to hear you express so much satisfaction at the Lists of Specimens forwarded by McFarlane and I. My collections at the Youcon are here now, but I have not yet examined them to see whether they be properly packed for the trip out next summer. I will do so however before spring.

My collection here last Spring was of no account. I think however, that I got a few good things among the Flycatchers and *Spizella pallida,* but probably they will have reached you ere this, having been forward by the boats of last Spring to Portage La Loche. I found this place much as I feared it would be—very inconvenient for collecting. Doubtless, as you say, there is much yet to be done here, but if mosquitoes are always as numerous as they were last Spring, it is utterly impossible to collect anything worth speaking of. Moreover at that season the Indians are continually arriving with their winter hunts, and have to be arranged with—the packing of the Furs, making up the post Accounts and other matters which require the constant supervision of the person in charge, deprives me of the time necessary for collecting and preserving any quantity. I have a clerk with me this winter, however, and I hope to be able to "go in and win, some good and rare ones"—"nous verrons". The Indians here are very much afraid to go any distance from the shores of the Lake in summer, on account of the "shaking bogs" which surround this place for a considerable distance.

I have never heretofore paid much attention to collecting anything but *Eggs,* but since birds skins, fossils, shells, etc. are all so much required, I will do all I can in that respect; especially the goose question. I shall endeavor to furnish you with a collection of all *[blotted],* that will enable you to set *it* at rest.

I received a small collection of Eggs from Lapierre's House this summer, but unfortunately, they were so carelessly packed, that a regular *smash* was the consequence. Mr. A. Flett also made a collection of 60 or 80 small birds nests and Eggs, but I have not yet had time to examine or label them.

I now forward a List of the Specimens forwarded by me last summer. You will see a number marked as coming from Mr. Brass, Fort Halket. I was much disappointed to find only one Egg in the collecting but, if I mistake not, it is a rare one—Franklin's Dusky Grouse. I hope he will do better next spring.

Wishing you all the compliments of the approaching season.

I remain, My Dear Professor Your's very sincerely,

J. LOCKHART

42. RODERICK MACFARLANE to SPENCER BAIRD

Fort Anderson 10 May 1864

[SIA, RU 7215, Box 14]

[Mr. McLeod "B" was Alexander McLeod, a labourer and a voyager in the Mackenzie River District. (H.B.C.A., PAM, B.200/z/1)]

My dear Professor Baird,

I have to acknowledge receipt of your kind and welcome favor of 17 October last, which reached me at Fort Good Hope, on the 19th ultimo.

Since I had last this pleasure in February, I have gone over, packed up, and forwarded (Case No. 9. Coll. '62) to Fort Good Hope, the "Anderson Collection" of "1863" (case No. 10. Coll. '62) together with the skeleton of *Ursus Arctous*? and *Rangifer Groenlandicus*? belonging to Collection "1862", which had been retained for transmission by the boats of this season. The whole, (as per Packing Account herewith) including the 2 cases (Nos. 3 and 4) left along the road last spring, amounts to 18 cases and cassettes, thus, 4 cases to collection '62, and 14 cases and cassettes to that of '63; and I sincerely trust that they will all reach you in safety, and their contents prove (tolerably) satisfactory.

When packing up the Collections, I found most of the eggs which had contained large embryos, so much broken up, that I could not possibly forward them; but I sincerely hope that the ensuing season will enable me, not only to represent most, if not all, of the species, with better specimens; but also to add a number of new eggs to the list. By comparing the birds and eggs collected last July on the Arctic Coast, and then entered as *Tringa Wilsonii*?, with undoubted specimens of the species, I have since been led to ascribe these to the *Ereunetes Petrificatus* of the *Report*. Having also obtained from the Esquimaux, lots of eggs and birds of what I believe to be the *Bernicla Brenta*, I think that a large number of *unidentified* eggs of "1862", really belonged to this species. *B. Nigricans*, I am not sure about, altho' its eggs are possibly represented in both collections? I am now also of opinion that the eggs entered under *Picoides Arcticus* (No. 570. Box 32. Cassettes No. 12) really belong to the *P. Hirsutus*. The latter is, to my certain knowledge, a bird of the "Anderson" *fauna*; and I think you may safely reckon on obtaining eggs of this species in course of the Season. At all events, you may rest assured that I'll do everything I can in the way of collecting, especially in regard to the Wax Wing. As to *Perisoreus Canadensis*, I am sorry to inform you that the contents of 2 nests; found by Indians last month, and containing 4 and 5 eggs respectively, got broken on the way in to the Fort. In this way also, I have already lost 4 eggs (I believe) of *Syrnium Cinereum*, and 6 of *Corvus Carnivorous*. But as I have now supplied several of the Indians with round Boxes, and also told them to be particularly careful for the future, I hope I'll not again have to record so many annoying losses. There has been no *thaw* as yet this spring—nothing to last year—and this will probably prevent the *Raptores* etc. from depositing their eggs, until the weather becomes finer?

I was absent at Fort Good Hope during the entire month of April, and this will account for the non-appearance of Mr. Whisky John in *this* year's collection.

I have lately heard from all the District officers, and nearly all of them *talk* of doing something in the egging way this Season. I trust that they will do so as, with the exception perhaps of Youcon, Anderson and Simpson, the other posts are still new and comparatively unexplored fields for operations. I have myself done nothing since my return from Good Hope; but I expect to do a little by and bye, and altho' I have no expectation of getting as many eggs of some species (*Lagopus Albus, Archibuteo Lagopus, Aquila Canadensis,* etc) as were obtained last year, I still hope to become again acquainted with the eggs of all these, as well as with that of many other Birds. Should I however fail, it won't be for want of trying at least.

In May '63, I informed Governor Dallas that I had forwarded and also secured two or three specimens of the Barren Ground Bear, for the Smithsonian; he answered under date 9 Decr. last as follows: "The specimens you allude to as having sent to the Smithsonian Institution will no doubt be acceptable; but I should prefer that you sent them to my address to the Hudson's Bay House London." This will make no difference in my disposal of the specimens. They were collected by me *for the Institution,* and I had already packed up, addressed and forwarded them to Good Hope, ere I received his letter. You will therefore receive these; and I will this Season endeavor to secure a specimen of each *Ursus Arctous?, Ovibos Moschatus* and *Rangifer Groenlandicus* expressly for his Excellency, and if successful, as I fully expect, I shall transmit them as directed by the boats of next Season. I shall also explain the matter to him, and also state that these specimens (I have sent) were absolutely necessary for comparison with the *Ursus Arctous* of Europe. I should like to send Governor Dallas a suit of my eggs, if you have no objections, I suppose that you would prefer me to do this thro' "the Institution", instead of *direct* from here. I do not however wish to put the Institution to any inconvenience on my account. I presume that 1 or 2 eggs of each species (not of the *rarest* of course, until they are better represented) I have collected will be amply sufficient. Please write to me on this subject. I should also wish to present Dr. Brewer with a similar suit of such eggs as may be wanting in his "collection", and of which I have already, or may hereafter collect specimens, as a slight acknowledgement for the pleasure I have derived from the 1st part of his able work on American *Oology.* By and bye, when my Collections will be much fuller than they now are, I may again trouble you in this way on behalf of one or more public Institutions. But if it be against either the custom or rules of the Institution to present objects of Natural History to individuals, then of course I must, and shall willingly conform thereto. I am afraid that you will think that I am beginning to give you much trouble.

As I have obtained permission to make another overland journey to the Arctic Sea this summer, I trust that I'll be able to get off by the 10 June. If so, and the weather etc., prove propitious, I think we will do a little in the egging way ere we return. I intend taking a small birch canoe along with us for the purpose of visiting the Islands alluded to in my previous letters; and my route thither will probably be much the same as last Summer. My party will consist of 10 or 12 persons, mostly Indians. I *must* however be back in time to proceed to Fort Simpson in the Autumn. I regret this, as I should have liked to secure specimens of the young birds of the year in the pure down and succeeding stages. I shall however do what I can in this line.

I'll not forget too, to tell the Anderson Esquimaux to be on the lookout for eggs etc., still I do not expect that the Collection of "1864" will be as large as that of "1863"; but I fully trust that, taken altogether (provided everything goes on well) it will be, at least, as valuable. *Nous verrons.*

I am so much pressed for time that I really cannot afford to address you at any length on the present occasion. I trust however to make amends, if necessary, by the letter I intend writing on my return from the Coast, and that I shall then have some interesting details to communicate. My absence (longer than I expected) at Good Hope has put me in arrear, so that the documents I had intended forwarding now, must remain until another opportunity. I trust also that any errors etc. that may be found in the list etc. will be ascribed to their true cause, ignorance or oversight. But to speak truly, I am, I assure you, far from pleased, not exactly with what I have done, but with the indifferent manner in which a large portion of it has been done. The great number of badly prepared specimens, etc., I cannot bear to think of; but I hope we will do better in that way for the future. I may also mention that I am now entirely out of *small* boxes (Esquimaux boxes I can always have), cotton, and paper for packing etc. I should also like to have 2 or 3 blowpipes and as many egg drills sent to me by *Winter packet,* together with a small case of dissecting instruments; and if possible, a supply of collodian. The Instruments I wish on private account.

I sent you a few Reindeer tongues last year, but do not know if they reached you in safety. I have since also ascertained that the case containing them was broken into by one of our men, and at least 1 dozen of them taken. I have now however packed up in Case No. 14 5 dozen cured Tongues to replace the missing ones. I trust that they will go down all right, and that you will do me the pleasure to accept of them.

In Case No. 11 you will find specimen of the Blue Fox mentioned in my February letter, together with 2 mountable and 1 skeleton specimens of *Vulpes Lagopus;* and I have just packed up in Cassette No. 13, 2 more mountable skins of the Blue Fox, received lately from an Esquimaux. This fox is very scarce. Among the post Returns, I have sent out but 6 skins this Season; of other Foxes however, I have traded considerably more than last year, especially Red, Cross and White. Black Bears are not scarce along the Lockhart River; when I first descended that stream in June 1857, we shot 2 out of 7 animals seen. The beginning of May, may be set down as the period when the *Lagopus Albus first* begins to assume its Winter Coat in this quarter. The changing invariably begins near the head of the bird. *Lagopus Rupestris* during the months of March and April at least, approaches us nearer than it does during the breeding season. It is then to be found within a few miles East and S.E. of the Fort.

My Golden Eagle is all right. He has been outside in his "Cage" since February last (end) and is now rather wild, and not at all to be trusted outside of its prison. I daresay it could fly well enough now, after very little practice. It is quite lively, and in general eats well, but neither in abundance nor voraciously. We have hitherto fed it upon Deer's meat (fresh) with an occasional diet of fish, which appeared to agree with it very well. I think it is a male. It has been very "restless" of late. If I only get off in time, and she have returned to breed in the same place, I am bound to get the eggs of this season, as well as the parent bird of these—of this eagle—and of the 2 eggs (285) I sent out last year. At all events, I have every intention of trying.

I have done nothing as yet in the Collecting line—this Spring—have got but 6 eggs *Corvus Carnivorous*?, which were found in a nest placed on a tall pine near this, on 30 April. By and bye, I trust we will do better. I have been busy writing letters etc. since my return to Good Hope, but as I shall have more leisure after the Packet leaves, I'll be on the look out for specimens. An Indian arrived y'day who informs me that he found a Golden Eagle's nest not far from here, but that no eggs have been deposited yet. He however saw both parents, he says, distinctly, and that the nest is situated on a tall tree! I'll soon find out all about this.

Cassette No. 13 contains Weather Registers for 12 months. I do not think they are very complete. They were taken by one of my men, (Mr. McLeod "B") whom I always leave in charge of the post during my absence. In the Cassette you will also find copy of my "NoteBook" from 15 May 1863 to 10 May 1864. A few specimens of Collections '63 still remain inland, while most of the Bird skins are very badly prepared, and the mammals also but indifferently well. It also contains numerous other specimens—Birds, Mammals and objects of Esquimaux manufacture.

It is time to conclude this uninteresting scrawl. I hope we will have some "good things" to communicate next July. Meantime, With much Esteem, Believe me,

Very faithfully Your's,

R. Macfarlane

P.S. I am not very *fortunate* this year—at least as yet. An Indian has just come in to say that a day or two ago he found the nest of a Golden Eagle, the eggs in which—2 in number—had evidently been eaten by a Fox! (saw the fox tracks in the Snow etc.). It was built against a steep bank, along the Upper Anderson. I sincerely hope that *this* will the very *last* of our *Serious losses* for this Season.

RMF

43. JAMES LOCKHART to SPENCER BAIRD

Portage La Loche 1 July 1864
[SIA, HBC Corr Coll, Folder 26]
[W.C. King was an apprentice clerk and clerk at Forts Resolution and Rae during the 1860s. (H.B.C.A., PAM, B.239/k/13)]

My Dear professor,

Your esteemed favor of 5 April reached me on my arrival here the 25 Ulto. I am glad that you were pleased with the few things I sent out last summer. I now enclose an Invoice or rather *List* of Specimens collected by me this Spring. There is nothing new to speak of. You will observe, however, that I have sent the Eggs of the White Crane, but unfortunately they have no parent with them. There is no doubt of their identity however, for there is no other Egg in this country like them, except those of the Grey Crane, and the last are so much smaller, that is impossible to mistake the one for the other. The Egg of the Grey Crane which I got at Salt River has been lost in some unaccountable way while being carried across the portage here. The remaining cases of my Youcon Collections are now sent out. I opened them at Slave Lake to see that they were in good order. I had no means at command to pack them better. I hope they may reach their destination in safety. The Guides have received strict orders to treat the cases of Eggs "cannily", and Mr. Hardisty has paid them well for their trouble between and Red River.

The *Sterna Caspia* nests late. I made several Expeditions in spring in search of their Eggs but did not succeed in procuring them. Mr. King, whom I left at Fort Resolution, promised to try and get them after we left there. I hope he will succeed. I put a small parcel for Kennicott in one of the Smithsonian cases, please hand it over to him.

Mr. Onion wishes me to say that he did not receive all his collections in time to send out this year. For one article he mentions having secured ten specimens of *Colymbus Adamsii.*

In consequence of our friendship for Mr. Kennicott, we are all very much interested in the new museum, which has arisen in Chicago under his auspices. It is the wish of all the officers in McKenzie's River, therefore, that it should get the preference, when the Smithsonian Institution has any duplicates of our specimens to dispose of.

Mr. Hardisty has every wish to collect and render whatever assistance he can in furthering the objects and views of the Smithsonian Institution, but unfortunately he has suffered so much from sore eyes all last winter and spring, that he could attend to nothing. He is now recovered, and promises great things for next year. I will write fully by winter packet. I remain, My Dear professor Yours very faithfully,

J. LOCKHART

P.S. Mr. Onion has forwarded to the Smithsonian Institution a small case of plants collected from all parts of McKenzies River.

44. DONALD GUNN to SPENCER BAIRD
Red River Settlement Augt 3rd 1864
[SIA, HBC Corr Coll, Folder 20]

My Dear Mr. Baird,

Your note of March 6th came to hand in the beginning of April. As I had nothing of Interest to communicate at the time of its receipt, I deffered writing until after my trip to Manitowaba Lake. On the 13th June I left here for that place. The two last years having been so very dry and this spring the same. I feared that in traveling from here to Shoal Lake a distance of forty miles, no water would be found by the way. Therefore, I went West along the Assiniboine River as far as the Settlement at Poplar point above long Lake a distance of from 60 to 70 Miles from here. There leaving the River I crossed the plain (12 miles to 15 miles wide) to the south end of the Lake, near the Mouth of White Mud River. I learned before I left the Settlement, on the Assiniboine, that some of the people there had been a few days before my arrival, at the Lake, and harried all the Gulls nests on the Island. However, after having traveled so far, I determined to go to the end. On arriving at the great Marsh bordering on the Lake we met some Indians, one of whom I hired to accompany me to the Islands in the Lake. In the afternoon we set out on foot to cross a part of the Marsh to where the Indians had left their dug outs. After a very unpleasant tramp for the most of an hour through reeds, water and sludge, to mid thigh at times, there were the Crafts!! two hollowed logs from ten to twelve feet long each. The Indian pushed out the shortest but the steadiest of the two and placed himself in it pointing to me to place myself in the other. This I did, after which we pushed on thro narrow channels resembling broad ditches bordered on each side with tall reeds, after some time, we came into a considerable opening, the wind blew stifly from the North. The waves were high and every wave as it struck my lop-sided apology for a canoe, poured in a few gallons of Muddy water. My craft became unmanageable perhaps, in some degree, owing to my stiffness being far advanced in the decline of life, as well as being unaccustomed to a craft of its small dimension and bad condition. After a long struggle with wind and waves we got into the reeds and made for the Tents where we arrived about middle of the night pretty well tired of our trip. The following day I supplied two Indians with ammunition and sent them off at the same time promising them great rewards if they would bring me Gull Eggs. The day became very stormy. The Indians returned in the Evening stating that their Crafts were incapable of passing over the Waves between the shore and the Island. They brought a few black-headed Gulls, and we shot some of them during the day, as they were passing on their foraging into the plains whither they went in great flocks to prey on the native grasshopper, which is very abundant this season. I sent off my two Indians the 2nd time. They returned at night stating that they had visited the Islands but found no Eggs on them. You can scarcely form an idea of the bitterness of My disappointment—from the fact that the Franklin Gull was there in great numbers. And that a week earlier I might have secured any number of their Eggs. And now after my long journey attended with some expense, and not to be able to procure a single one was truly mortifying in the extreme. If you be still desirous of obtaining the Eggs of the Laughing Gull, and of the Franklin Gull, they being the only Gulls which I saw there, the best way to obtain them will be to leave here in the last week of May or in the first week of June. Two Carts drawn by two oxen or two horses, and two men and a lad, would be required for the space of three weeks, the two men to paddle a boat and hunt the lad to take care of the animals—while the former are hunting. They would require to take a light skiff from here, take Shoal Lake on their way out, cross from it to Manitowa Va-go along that Lake from thirty to forty miles to where the Islands are, where the gulls breed. After passing a week there and hunting for all that can be found on the Islands and in the marsh, retrace their steps by Shoal Lake, hunting for all such nests as may be found by the way. This spring I employed some boys to hunts for small bird nests. They brought me a few, with the parent birds which I have sent, but I fear that there is nothing of much Interest among them. The Grasshoppers made their appearance in the upper part of this Settlement in the last week of June. An immense swarm of them invaded the Settlement on the Assiniboine over a distance of twenty four miles. On the 6th of July, driven by a south-westerly Wind, they invaded the lower part of the Settlement. The Sky being clear and the sun shining brightly—short time before 10 a.m. We perceived them passing between us and the sun just resembling a thick shower of soft flaky snow. Some of them, some thousands of feet high, others very near the Earth. Tens of Thousands of them passed over to the East, and thousands of them fell on our fields. They continued passing for several days but fewer falling to the ground, than fell the first day. They committed great ravages on the Barley Cutting of the heads. They attacked the leaves on the Wheat and Cut them of, at the same time cutting of many of those that had lately made their appearance, the same being soft. On the 11th they were observed pairing and laying their Eggs in which work they continued until the latter end of the Month. The greatest Coupling and laying from the 15th to the 25th. By the first of this Month they were done and they have now nearly all disappeared, but we have not seen them take their departure. Therefore I cannot say in which direction they went. Many of them have died. Some I have seen with a hole on the Thorax with a stout white magot attached to the lip of the hole. I have sent two vials containing grasshoppers. In one of the vials there are a number of these creatures taken in the act of copulation. Those in the other vial were taken singly, in a third vial you will find a lot of their Eggs and I have counted the Eggs in each of several nests and found them to number from 25 to 31. The first being lowest number the latter the highest found in any. I have been looking forward for some of Your *reports* the last I received was that of /61. I shall be most happy to have the series. The box containing specimens I sent to Upper Fort Garry last week. Mr. Mactavish promised to forward it by the first opportunity.

Believe me, Dear Sir, Yours Sincerely

D. GUNN SERVT

45. RODERICK MACFARLANE to ROBERT KENNICOTT

Fort Good Hope 9 Sept. 1864

[SIA, RU 7215, Box 14]

[William James McLean (1842-1929) was an apprentice clerk at Fort Norman in 1863-64, and the clerk in charge at Fort Liard from 1864 to 1873. He is, however, better remembered for his role as the Chief Trader in charge of Fort Pitt during the North West Rebellion of 1885. James MacDougall (1843-1904), a native of Scotland, joined the HBC in 1862 as an apprentice clerk. He was immediately posted to Fort Yukon, and stayed there until 1869. He eventually became a Chief Factor and an Inspecting Chief Factor in 1879 and 1889 respectively. Alexis L'Esperance and Baptiste Bruce were responsible for the Portage La Loche Brigades, which carried freight from the Red River Settlement and Norway House to the Athabasca and Mackenzie River Districts. L'Esperance had headed up the brigade from 1833; Bruce joined him in 1848. Bishop Vital-Justin Grandin (1829-1902) left France for Red River in 1854. In 1857 Grandin was chosen to assist Taché in establishing the Oblate missions in the northwest, and he therefore toured the Mackenzie River District from 1860 to 1864. Henri Faraud (1823-90), a French Oblate, established the first Roman Catholic mission in the Athabasca District at Fort Chipewyan in 1849, and another at Fort Resolution in 1852. Faraud was placed in charge of the vicariate apostolic of Athabasca-Mackenzie created by Taché in 1863. Pierre-Henri Grollier (1826-64) was Faraud's companion during the 1850s. Brother Joseph-Patrick Kearney (1834-1918), originally from Ireland, was sent to the district in 1857. In 1861 he was posted to Fort Good Hope where he stayed until his death. Kearney shared the Good Hope mission with Jean Séguin (1833-1902), who stayed at Good Hope until 1901. In 1862, Father Emile Fortune Petitot (1838-1917) was also sent to the north as part of Taché's missionization program. He remained in the Mackenzie River District until 1882, and during that time he undertook geographical explorations as well as working for the church. Petitot's studies of indigenous languages and folklore were published, as well as some travel accounts and ethnographic monographs. See, for example, *Tradition indiennes du Canada Nord-ouest* (1887); *Dictionnaire de la Langue Dene-Dindjie* (1876); and *Les grands Esquimaux* (1887). Pierre Allard was a fisherman in the Mackenzie River District who retired to Canada in 1866. (H.B.C.A., PAM, B.200/z/1)]

My Dear Kennicott,

I have delayed writing to you until the latest possible moment, and the consequence is, that I must be very brief in my reply to your own long and interesting communication of 15 April last. I shall however write you more fully by Winter packet.

Should your own and Professor Baird's future letters to me prove as flattering as those I have been accustomed to receive of late, I really must believe, what I have not hitherto done, that *I am* doing something in the way of advancing the interests of Science, however, as I myself experience much pleasure in Collecting objects of Natural History, I shall continue the occupation, equally regardless of praise as of censure. As for writing anything for *publication,* I'll know more of the subject than I do now, ere I be persuaded to attempt anything of the kind. As to the brief notes accompanying the specimens, I care not what use may be made of them, as they are, I think correct in the *little* they say. But enough of this for the present.

By the way, the guns and other goods left at Norway House in 1862, are still there! also the whole of the cases etc. sent up from Chicago and Washington last spring. What think you of this? How is it possible for us to go in strong in the collecting line, when the necessary preservatives etc. are thus detained along the road? If the Company and Governor Dallas (also) wish to take Credit to themselves for what they do for Science, I think they might first to *deserve* it; they can easily find room for the immense quantity (65 pieces this year!) of Mission property which annually encumber the Portage Brigades, tho' a great fuss is always made about the comparatively few pieces of property brought in belonging to their own officers. Governor Dallas, notwithstanding his assurances to Prof. Baird, is a *humbug.* Sir George Simpson never could have acted as he has regarding the Smithsonian Cases. *He* would have seen that *all* the outward and inward property was forwarded with the least possible delay to its destination. You know what Dallas told you at the portage—that we were to have each 3 pieces *free* freight allowed us annually, *over and above* any cases from the Institution—well, that promise was made in '62 and it is now '64, and it has not been fulfilled yet. *Au contraire,* by the Resolve 98 of this Season's Council, he has directed us to be charged freight on *all* property embarked in the Boats to our address, not only to the Portage, but to winter quarters! What think you of this? I could mention many other annoyances inflicted on us by our new Governor, who is I believe a pleasant enough man to those with whom he associates. In order that you may judge of his unpopularity in the MacKenzie River District, I may mention that McLean last fall proposed a bumper to his health, and that not a soul of the 15 officers their present would join him in doing honor to his Excellency. It is quite possible that the HB Coy may one day find that they have gone too far in their bad treatment of those who now hold the "mainstay" of the fur trade, I may say in their own hands.

It is quite needless for the Institution to be at the trouble of sending us Books, etc. if they are to remain a year or two along the road, and it would be as well to make Governor Dallas aware of the fact; if he pleads want of room in the boats, don't credit him—you as well as I, know that one or 2 pieces extra p. Boat won't make the slightest difference to the crews—besides, they (the Co.) never make any difficulty about bringing in any quantity of Mission property. Excuse this long tirade, but you will, I think, admit that we have had reason to feel somewhat *riled* on the subject.

As I have written very fully to Professor Baird anent this Season's Collections, I need not here trouble you with a repitition of the same. I may however state that I should have been better pleased if the Collection had encluded a few more of the many good things in the Oological line, still remaining to be added to the Anderson fauna. Perhaps we will be enabled to make a few more additions thereto next summer. I should like to get a supply of *fancy* articles for the egg trade. I can never have a sufficiency of these at Fort Simpson; and I should also like to get a good Maynard Rifle with a spare shot Barrel, and a supply of Ball and shot Cartridges. The price of these articles, I might either send by Draft, or disburse among others in here on a/c of the Institution—just as you think best. Jones also desired me to mention his wish to have a shot Barrel sent to him. Gaudet too would like to have a Rifle sent to him on his private account. I am extremely obliged to you for the trouble you took about the articles I requested in '62. I'll surely get them next year—the Governor won't detain them another year. I shall find it no easy matter to pack up my bird and eggs of this season, for want of paper and cotton. I have got a few newspapers for the former but none whatever of the latter article. Last year's collection ought to reach you in good order and condition, as I got the whole as well packed up as I possibly could, while all the cases containing eggs were entrusted to the most careful of our Steersmen. I paid them for their trouble. I am also indebted to Mr. Hardisty for having paid Bruce and L'esperance to carry these, as well as to look after the other cases, on the way down to Norway House. I wished to pay them myself, but Mr. H. would not hear of it. In Augt '62, I requested you to send me *Harper's Weekly* and *Harper's Magazine;* but the letter, I think, never reached you. I should now however like to have them sent me regularly. I want them on my a/c. I shall send in a Draft for these and the other articles next fall after I receive the Invoice.

Lockhart got some excellent things at Resolution this Season, among others, the eggs of the White Crane (*Grus Americanus*). It is strange we cannot, very often, find eggs of the *Colymbus Adamsii.* MacDougall got a few nests of owls etc. at Youcon, but he was absent at La Pierre's House during the best time for collecting, or he would have done much better. Sibbiston however made a good collection, so Jones informed me. James Flett did very well at LaPierre's House. Gaudet and A. Flett got nothing, especially the latter, to speak about, but all of them promise to do something next Season.

Jones was at Simpson this fall—the first time since he entered the District in 1859—he will remain inland next Summer. MacDougall will however be coming up to transact the Youcon fall business. There have been no changes in the post appointments this fall; all the officers have gone back to their quarters of last winter. Lockhart however expects to go out next Summer; should he do so, he will probably be succeeded by Macdougall.

We have been very dull this fall, here, for want of something of an elevating tendency. Couldn't you send us some; annually in the Cases? I see no necessity for asking the Governor's permission; he'll not alter his mind, for he is as obstinate as a mule on nearly every subject connect with "R" District. Why let him know aught of the matter? He tried to put a stop to *grog* all over the country, but was obliged to retract his orders to that effect, ere he had been six months in his Governorship. He'll never find out that *you* send us any, provided you pack it up carefully, and apart from the Smithsonian publications etc. Alcohol for preserving might be sent without exciting suspicion, but it is not good for drinking. Jones had in some superb whiskey from Canada this fall. I make no difficulty about asking for a supply of the "water of life", and I sincerely trust that *you* too won't at any time hesitate about asking *me* to do or procure anything for you that you may require from this quarter.

Taylor got married to an American Lady this fall. I believe she is a Native of Maine, and was formerly stewardess of the *International* Steamer. I think she and Taylor will get on very well together. She was accompanied by a grown up daughter, and they all pass the winter at the new establishment on Great Bear Lake.

Bishop Grandin has gone to Isle a la Crosse, where he will now always be stationed. Bishop Faraud will be coming to the "Rapid" next fall. Poor Grollier died here on 4 June last. Seguin is now in charge of this Mission, and a new Priest, a fine clever man of the name of Petitot (a frenchman) acts as his coadjutor. Kearney is also here.

Most of the old hands among the men, whom you knew, are still in the District. *Allard* has been twice at the Coast with me, and a good hand I have found him for Collecting, I assure you. He has not however lost aught by the great interest he always displayed on these trips. I having well recompensed him for his trouble. I expect to go to the shores of Franklin Bay again next Summer; but as I must be at Simpson in the Autumn, I shall probably, as was the case last summer, have to quit the coast too soon for the nesting of the Eiders and other Sea Birds. However, I shall do what I possibly can, you may rest assured.

I need not assure you that the Chicago Museum is perfectly welcome to a share of my part and future Collections in Natural History. I shall try hard to get a Bear and Muskox for you *next* Summer. The Specimens received this Summer have all been promised, as you should have had one of them. I spoke to J. Flett about the sheep and goats, of the Rocky Mountains, and he'll doubtless try his best to oblige you. I have also directed the attention of both Fletts to the importance of securing specimens of the grey Bear of the Mountains, for comparison with the Barren Ground species.

If I remain a couple of years more in this District, I shall try and secure a *live* Musk ox for the Smithsonian, tho' how to get it taken out in the Boats, would be, I fear, a rather serious affair. *Nous verrons.* We got a young one last Summer, but it unfortunately died. I have now got 4 living Eagles (*A. Canadensis*) 1 from last and 3 from this summer, the former stood the winter very well, and I have no doubt the others will do so equally well this Season.

I shall write you more fully by Winter packet. I shall then also answer your letter. I cannot at present do so, as I am much pressed for time. I start tomorrow for Winter quarters. (Please excuse this very disreputably written scrawl.)

I have got no labels to attach to my specimens I may collect hereafter, in fact, many of those tied to last Summer's specimens, are mere pieces of paper bearing the simple number thereon. Could I not have a large supply, say 2,000!, at once? and thereof save me future trouble. For the future also, I have to request, that any Books from the Smithsonian, or Articles on my private account, be addressed direct to myself. It is a mistake to suppose that those being addressed to the Gentleman in charge of the District, will ensure their prompt transmission from Red River. I thought that your own experience of the plan, proposed and adopted by Mr. Ross, had disabused your mind on the subject of the *outsiders* making any particular difference between the property of the head and the subordinate officers of the District. Witness again this year, the cases have all stuck, altho' addressed to Mr. Hardisty.

I must now conclude this scrawl. Please excuse it. And with best wishes for your health, success and felicity. Believe me, Ever, Very faithfully your's

R. MACFARLANE

46. JAMES LOCKHART to ROBERT KENNICOTT

Fort Resolution 21 November 1864

[SIA, HBC Corr Coll, Folder 26]

[*Che-tsoh-kah-kieh* was the name given Kennicott by a Kutchin Chief. *Che-tsoh* means bird, and *kah-kieh* means chief, but as Kennicott pointed out: "... they usually drop the *kah-kieh* as it is the appellation of any gentleman, as well as of their own chiefs, and call me *Che-tsoh*." ("Journal of Robert Kennicott," in James, *First Scientific Exploration*, p. 85.) Mr. George C. Walker of Chicago aided in establishing a museum at the Chicago Academy of Sciences where Kennicott was given a curatorship in 1864.]

My Dear Cheeh-tsoh

The time approaches for our usual winter packet to leave the District, and I know you will be looking for a long "yarn" from my pen. You will, however prepare yourself for partial disappointment. It is with the utmost difficulty that I have "screwed" myself up to the writing point. You know well my habits of procrastination and "constitutional disinclination" for writing; added to that I have been ill more or less ever since my arrival here in September and am still unwell: to crown all, my right hand is so swelled up and painful from a "whitlow" or something of that kind that it is with difficulty I can hold the pen. I am consequently in not the most amiable mind imaginable and can only sympathise with "the bear with a sore paw"! I feared at one time that I would be unable to write a line to anyone, but thank God I can still shew my friends that I bear them in kindly remembrance.

I wrote to you in summer from portage La Loche—that letter I suppose you have received ere this. From the portage we had a tolerably quick trip down to Fort Simpson, where I had the pleasure of meeting all our old friends. Jones from the Youcon, the two Fletts from Lapierre's House and Peel's River, McFarlane from the Anderson, Gaudet from Good Hope and Brass from Halkett, all were well and we conversed much of you. We had several very pleasant "reunions" and your health was proposed and drunk with all the honors. Jones was the luckiest man among us, for some one of his friends in Canada sent in a Case to him containing a tin can of 6 gallons splendid whiskey—on the "sly" of course. The case was packed tight with sawdust round the can; the latter full to the cork and then soldered airtight. A capital plan, which I would recommend to be adopted by any generous philanthropist, who may feel inclined to benefit and succor his fellow creatures by sending some medicine to the poor exiles of McKenzies River. Macfarlane and I exhorted them all to collect specimens for you and the Smithsonian and they have all promised to do what they can. Brass tells me that he is unfortunately always away from Halkett before the birdnesting season commences, and you know that there is nobody left there except an indian in summer to make the hay. James Flett has given me a case for you containing two fine specimens, one of the goat in winter coat, the other a young female Carribou, also a small box of Eggs. The case is well packed for I have examined it, so I will, in spring, just send it on as it is.

Andrew Flett made no collection I believe, but I told that your orders were that only in the event of his making a good collection, would he get the gun that was coming in for him. He was very sick all spring he says. I heard that they had made some collection at the Youcon too, but I did not hear what it was.

Is it not strange that none of the things of which you sent me Lists from the portage the year you went out, have made their appearance as yet. Neither have I heard anything of those sent last year. I thought that perhaps the Athabasca boats would have brought them on, but it appears they have not. Since my arrival here I received a packet of photographs the annotations and address in your never-to-be-forgotten, or mistaken, fist for which mark of attention I return you my most hearty thanks.

My Dear fellow, there is certainly a fatality attending my small endeavors to continue my collection here. You know that in my last epistle I mentioned the almost certainty of Mr. King being able to procure the Eggs of *Sterna Caspia* in summer. Unfortunately, however, soon after my departure for the portage, both of his men and all the women fell ill. One woman became as mad as a March hare, and he had to mount guard over her day and night himself. She "died" and became as stiff as a poker and quite cold twice, but came to life again some hours after. She has now recovered her reason and is nearly well. With all these "malheurs" it was impossible for him to leave the Fort.

I have collected a lot of shells from Little Buffalo River, also a geological specimens from around Great Slave Lake. Hang the drawings! I can make nothing of them "J'ai perdu le tour tout à fait." I send you one nice one of Fort Rae done for me by the priest Père Pettitot, also one or two "botches". Confound Mr. Ross for walking off with all my sketches. I would like to send you a sketch of my train of dogs, but I "can't come it". I'm sure I will meet with your cordial sympathy when I inform you of my great loss: my splendid foregoer!, my "Royal"! The most sensible and affectionate dog that ever walked or ran on four feet; infinitely more sensible than many who go on two! He died during the summer which I was off to Portage La Loche. Mr. King got him skinned and stuffed for me and his remains were interred with due solemnity in a prominent spot at the foot of the garden near the ice house. Peace to his remains poor fellow! Upon my honor I felt his loss much the same I think, as one would feel to lose a child! My Youcon train of dogs are at Fort Simpson. Mr. Hardisty asked them of me, so I gave them to Mrs. Hardisty to drive about with. "London" and "Caesar" are as smart as ever yet. I'm sure you sometimes, in your dull smoky cities, "think long" for the free wilds of the North, for a good train of dogs and an open track thro' the thick woods, over the icebound lakes and the snowy prairies. The weather not too cold, and the dogbells ringing cheerily, to some câche of good fat moose meat. How jolly one feels in the camp seated before the blazing fire with a pipe in one's cheek after several pounds of fat Rib or "broehaet" and five or six pots of strong tea have been safely stowed away under ones belt, discussing the comparative merits of all the dogs, men and women and things in the district. I place "dogs" *first* you will notice, for the very sufficient reason to me, that they stand highest in my estimation. Many would say that is rather an uncharitable estimate surely, but *you* know, that the halfbreeds and others up here have a thousand bad qualities and failings from which the poor dumb brutes are free, while these last have many virtues, of which the others are just as innocent.

I must now tell you that our prospects of passing a comfortable winter as regards "grub", are not particularly bright. The fall fisheries all proved total failures. In consequence we will be obliged to keep the Nets and Trout lines under ice all winter, and you know that is no joke on Great Slave Lake in January and February. The provision Trade at "Lockhart's House" or old Fort Reliance was tolerably good but not equal to last year, altho' I sent Mr. King there this fall to oversee the business. Neither will we get so much fresh Meat from our hunters this year. Last winter the snow was so deep that the moose and deer were run down on Snowshoes and killed with arrows with the greatest ease. The consequence was that in Spring there was not one animal left alive with some days march of the Fort, and there are very few moose tracks to be seen this fall. We have only received 6 moose from our three hunters as yet. The season has been extraordinarily late this year; 'tis only three days since that the lake froze so that we could put down our nets under ice at "Round Island". Reid made a splendid fishery at Big Island this fall, so Fort Simpson will be abundantly supplied with fish for the winter. By the bye, old Taylor is married, to an old Yankee lady, who came in last summer. The old fellow seemed quite happy, and soon learnt to walk arm in arm with his lady. She was a dressmaker or something of that, not very highly educated I should say, but quite good enough for old Taylor. At portage La Loche, she used to make Dunlop blush with her, to us, strange modes of expression. Enquiring after the health of his lady of a morning, she would say, "Wall mister, how's your *woman* today?" Taylor has shifted his Fort again, and is now settled on the shores of Great Bear Lake. The old place is abandoned altogether now. Macfarlane was delighted you may be sure that the Anderson is to get two years trial more, he will therefore have plenty time to finish his collection of the Eggs etc. of that place.

Mr. King remained here this summer while I went down to Fort Simpson with the boats. On my return he went off to Fond du Lac to have some fun among the deer, he only killed two or three however. He then went off down to Fort Simpson by the fall boat, after the Athabasca letters arrived. He will return here I expect, with the December packet. Mr. Reid will come over here with him to spend a few days here. If my health permit, I shall go over to Fort Rae in February to meet Mr. Onion who is coming up to meet the Spring packet. Smith will then come over here with us for the same purpose, so we hope to have pleasant times of it. I regrett however that I shall not be able to offer them a "horn"; but that can't be helped.

I hope to get Furlough next summer, but I am not at all certain about it. I think I told you in any last letter that I had heard very bad news from home. By the Athabasca boats I received another letter from my brother. My Father was then unable to write still, but was no worse than he had been, he said however, that if I did not go down till 1866, I would never see him. He seemed to think he had not long to live. In any case when I do go down, I shall try to hunt you up, if the Yankees will only let me. I wish to goodness you would "kiss and make friends", with the South again. How did you escape the Draft? I thought no one was to be exempt except parsons. If all is true that

Harpers weekly says, I suppose the crowning campaign is about at a close and the War ended by this time. I hope it is with all my heart.

Your Chicago "sucking Smithsonian" was brought on the "tapis" and discussed in Council at Fort Simpson this fall. The general wish expressed was; that you deserved all the aid we could give; that we should continue to send any Collections we might make, to the Smithsonian as usual, with the understanding that in the distribution of duplicates, your Institute should have the preference over all others, home or foreign.

I must write to your friend Mr. Walker, but what I can write that will interest him I am sure I cannot tell. I wish you were here to prompt me.

I think somebody must have lit his pipe with my Louchoux Songs for search as I may, I cannot lay my hands on them. What a careless fellow I am about my papers. I lost my Commission even and did not find it for two years! This letter has been written with great difficulty and by fits and starts, which will fully account for its prosiness and disconnection. My hand is so painful that I cannot write for two minutes in a stretch. Wishing you every happiness. I remain, My Dear Kennicott, Your's ever the same,

J. LOCKHART

47. JAMES LOCKHART to SPENCER BAIRD

Fort Resolution Great Slave Lake 28 November 1864
[SIA, HBC Corr Coll, Folder 26]

[Jefferson Davis (1808-89) had been both a congressman and a senator before becoming a secessionist and the president of the Confederate States of America in 1861.]

My Dear Professor,

I have done so little towards increasing the general stock of knowledge regarding the Northern Fauna during the last two years, that at last I am almost ashamed to write to you. I have, however, once or twice taken occasion to mention to you and Kennicott some of the obstacles which present themselves in the way of the Collector at this place, and trust, therefore that you will not attribute it to laziness or carelessness on my part. I had hoped to have sent out a good collection from here this summer. I had spoken, preached and promised so much to the "Yellow Knives" from the Mountain-Country and along the Barren Grounds at the Northern end of the Lake, that I felt confident they would give themselves some trouble to bring me specimens, for 'tis from there I have the only chance of procuring some rarities not yet received, such as Eggs of *Colymbus Adamsii, Lagopus Rupestris, Cygnus Amer.,* Threetoed Woodpeckers, Crossbills, *parus* and others. All my speechifying was vain. I received nothing. At Salt River, I engaged one of the Beaulieu's to collect but he and all his family unfortunately fell ill and could do little or nothing. This year I have engaged him again, and he has promised to procure for me a strongwood Buffalo and a Carriboo, besides White Cranes' Eggs and other things which will please you doubtless, and can only be got from there.

At Fort Simpson, I received a Box from Mr. James Flett of Lapierre's House, containing two fine skins, apparently well preserved, one of the Mountain Goat, the other of a young female Carriboo, both of them in their winter coats. These I know you will be glad to receive. There is also a small box of Eggs and a few other specimens in it.

I am glad to hear you say, that some of the few specimens forwarded, from here, will be of use to you for mounting, and that they are to figure in detail in your new work on Arctic Fauna. I may say, however, that heretofore, I have never given myself much trouble to collect mere skins, my attention having been principally directed towards "Egging": but since I find that I can do so little in that way now, I shall employ myself with the birds and beasts, whenever I can lay my hands on them.

I was unfortunate, last spring, in not being able to get any specimens of the *Anser Rossii.* Very few passed this way. I did not even get a shot at them. I suppose they must have gone North by the same route which they generally take on their return South in fall, for notwithstanding the great numbers which usually pass here in Spring few or none of this species are ever seen returning past here in fall with the other wildfowl. *Colymbus Adamsii* is another species which we rarely see in fall, tho' numerous in Spring.

On account of our friendship with Mr. Kennicott, we all feel an interest in his new Chicago Institution, and have every desire to aid him, and contribute to its success, with our collections. We will, however continue to send *all* to the Smithsonian as usual, with the request, that in the distribution of duplicates from thence, Kennicott may have the preference.

It will not be difficult for me to procure any quantity of the Skins of *Sterna Caspia,* for they are numerous here in Spring. I had hoped also to have been able to tell you that I had its Eggs, but I have been disappointed—better luck next time perhaps. I have never seen but one Godwit here, that is the one I sent you last summer. They are very rare indeed.

I have applied for leave of absence to go to Canada. Should it be granted, it is possible I may have the pleasure of seeing you, for I shall assuredly go to see the Smithsonian Institution. That is to say, if the Yankees will allow me. I sincerely hope that by this time that dreadful war is at an end, and Jeff Davis made a slave of, in his turn.

Last month I received a jolly lot of photographs. The parcel came in by the Athabasca boats unaccompanied by any letter, but I knew Bob Kennicotts unmistakable fist in the address.

We have had most extraordinarily mild weather this fall; the like has not been seen for many years. Not one cold day have we had yet and little or no snow.

Hoping to hear per Spring packet of the safe arrival of last summer's specimens at the Smithsonian. I remain, Very truly yours'

J. LOCKHART

48. JAMES LOCKHART to ROBERT KENNICOTT

Fort Resolution Great Slave Lake 5 December 1864
[SIA, HBC Corr Coll, Folder 26]

[In this letter Lockhart attempts to demonstrate his worldliness by discussing the American Civil War using the epithet "nigger drivers" to describe the Confederate States. Lockhart appears to be anti-slavery while another northerner, Lawrence Clarke, supported the southerners. See Clarke to Kennicott, 16 Jan. 1865. François Beaulieu (1771-1872) was an interpreter in the Athabasca District, 1859-70. In the 1867-68 outfit he was also the Post Master at Salt River. Olivier Laferté was a steersman in the Mackenzie River District. In 1864, he retired to Red River after having been present at the scene of Pierre Gendron's demise near Fort Rae. Gendron was allegedly murdered by William Thomson Smith, a HBC employee.]

My Dear Cheehtsoh,

You tell me in one of your letters that you occasionally send me *Yankee Notions* and *Harpers Weekly.* The latter I receive, but many of the numbers are wanting. The former I have never seen at all. Of *Wilkes Spirits of the Times* sent by Mr. Walker's brother, I have never seen a single number either, so you see there must be some most unscrupulous rascals, whether British or American, along the postal route between here and Washington. The *Yankee Notions*, I regret exceedingly, for doubtless they would furnish matters for mirth, and you know one "feels so good" to have a jolly laugh on a cold day.

What think you of the Eggs of the White Crane? I hope no American will have the audacity to doubt their identity. If they do, I'll go down on purpose to pitch into you all right and left! What would you do then, I would like to know? With me on the North and the "nigger drivers" with Jefferson Davis at their head, on the South, I guess we'd soon wipe you out and end the War! Nevertheless I'll first try a more peaceable mode of establishing their identity (not from any fear of getting my own head broken, be it remarked). I have again engaged Francois Beaulieu to collect for me at Salt River, and I hope thro' his instrumentality, should we be alive and well, to send out a lot of these eggs identified.

Didn't a gentleman at the Smithsonian who is working up the Gulls or some other Birds, write to me recommending me to send lots of Birds Skins and specifying the particular species to which to turn my attention and time? Such an impression is on my mind, but I cannot lay my hand on the letter, neither can I recall his name to my memory. If such a letter was sent, you may assure that if well, I will collect everything I possibly can, Birds Beasts, Eggs etc. but damn the Fish, I can't do them. I have made a "mull" of several already and pitched them to the dogs!

But here I am running off into all manner of subjects when my object is now sitting down to write to you, was to tell you a story, I heard a few days ago, of how the Coppermine River was first discovered and named.

In times past, the Yellow Knives and Esquimaux were always at

War. On one occasion the Yellow Knives having followed the Deer far over the Barren Grounds and committed great Slaughter among them, were encamped on the border of a small lake drying their meat over fires made of moss, and by the heat of the sun, preparatory to their return to the wooded country. Having seen no traces of Esquimaux during their peregrinations they reposed at their ease, unguarded and unwatchful. They had been discovered however, by a band of Esquimaux who one night made a descent of the Yellow Knife camp. Several were killed, but finally the Esquimaux were repulsed and they again disappeared in the surrounding darkness. In the confusion which succeeded, it was not discovered that a woman was missing. On the following morning the dead and wounded were attended to and the poor woman discovered to be missing. Their first thoughts were that she had sought safety in flight during the fight, and was sure to return to the camp to see the result of the fight, and in hopes of finding some of her friends who might like herself have sought safety in flight. The day passed and no one appeared. It was only then that the certainty presented itself to their minds that she had been carried off by Esqx. but now pursuit was useless, so they packed up their meat and "train" and returned towards their own country. In the meanwhile the woman was on her way, hurried along blindfolded amidst the band of Esquimaux towards the sea. She endeavored frequently to leave traces of their flight, but the watchful Esqx. always took care to obliterate any sign she made. She tried to obtain some glimpses "en cachette" of the route they were travelling, but the lynx eyes around here were not to be so deceived, so she was threatened with instant death, should she attempt to remove the bandages from over her eyes. In this manner they travelled, it appeared to her many suns, taking little rest. At length she found herself placed in a boat made of Skin and for several days they continued their journey, only putting ashore at night. When her eyes were unbandaged she found herself among a large number of Esquimaux lodges, situated on a long narrow sandy point stretching out into the sea, she had not, however, the slightest notion which direction she had come from. She was immediately taken to wife by one of the principal men of the band, and being strictly and constantly watched, she saw it was certain death to attempt an escape. Two years thus passed and she had a son who was beginning to walk about. She had frequent opportunity of escaping during that period, but while her baby was small, maternal affection prevented her from hazarding its life with her own, for she was still ignorant of the direction in which her own country lay, and she thought, that by delaying the attempt and dissembling so as to appear quite contented with her lot, she might in some unguarded moment, succeed in discovering it. At last she determined to make her escape, or die in the attempt. So one night, after all the men had started on a short hunting excursion, she armed herself with two staves, one in each hand, to steady and support herself, (as is their fashion when carrying heavy burdens), put her sleeping child on her back, and crept along the edge of the water, to pass the row of lodges, for the point was exceedingly narrow. Some of the dogs barked, but they had all been tied up to prevent them from following their masters that day, so she succeeded in passing all the lodges without alarming the inmates, and started off at a quick pace along the naked point of sand. It being quite dark no one could perceive her at ten yards distance. Two hours walk brought her to the extreme point, and there was nothing to be seen but water all around. She saw there was no escape for her, for she thought to herself, "should I return I am sure to be seen, and they will know that I attempted to escape, so they will kill me!" Almost heart-broken she sat down covered her face with her hands and wept. While she was thus lamenting her desolate condition, and in momentary expectation of hearing the Esquimaux in pursuit she became sensible of something licking her hands, which were still held before her face. On looking down to see what it was, she perceived it was a large wolf, who immediately trolled off from the point into the water. The woman looked after him and thought it strange that he did not sink above a couple of inches. In the meanwhile the wolf had returned again close to the woman, whined, and then walked off again. It immediately struck her that possibly this might be some road of escape, so up she got and walked after the wolf which seemed delighted to see the woman following him, so on he went, stopping occasionally to wait for her, and whenever she stopped to rest a minute, the wolf would immediately stop and sit down, till the woman again started. In this way she continued advancing all night and the following day. The pangs of hunger at length admonished her that she had forgotten to take any provisions with her, she nevertheless determined to follow her self constituted guide, the wolf, as long as she was able. At night sleep began to overpower her, so she stuck her staves crosswise into the sand as deep as she could, then leaning across them slept for a time. At length she awoke refreshed and proceeded on after the wolf, who had waited for her while she slept. The following evening, she arrived on a small island, on which were a few willows. She immediately made a fire, for both herself and child were benumbed with cold, they then lay down and slept. How long she slept she knew not, but when she awoke she perceived a dark mass of objects approaching from the direction in which she had come. She at once thought it was the whole tribe of Esquimaux in pursuit of her, and anticipating instant death, she threw herself down on the ground in despair, seeing that it useless to attempt escape from the small Island unperceived. Sometime after, on taking another look, while she clasped her starving child in her arms, to see if her enemies were near, lo, and behold! It was a large herd of carriboo. On they came in countless numbers, soon they were passing on each side of her and the little Island was covered with them; crushing and crowding to get a few mouthfuls of the moss that grew on the rocks around her. She succeeded in killing a young deer with her staff, before the herd left the Island on their never ending peregrinations. Her deer was soon dressed. She took the now [*cow?*], cleaned, it then filled it with the heart kidneys and other "tit bits", all cut up in small pieces, blood and grease were added and it was hung up to roast before the fire, by a string. During this process, she and her son appeased their hunger, by throwing pieces of meat on the hot coals and eating them half raw. Of the thighs and shoulders, she determined to make a provision for her further journey so they were immediately cut in strips and spread over the fire in the smoke to dry. She then replenished her fire and lay down to sleep. When she awoke the sun was again high, her roast well cooked, so she took it down, and they ate. She began to think at last that it was high time to resume her journey, and began to pack up her provisions. When ready she called her boy, but he would not listen to her, but kept eating away. The woman saw more immense bands of carriboo coming and fearing the Esquimaux would follow them, she determined to abandon the gluttonous youngster, knowing he was safe in their hands and he embarrassed her much in her flight. Away she went in the path of the Deer. The wolf again taking the lead, the moment he saw the woman start. The third day from there, having taken a short sleep occasionally leaning over her staves plant as before described, she saw a long low dark line along the horizon. Ere night she reached it, and found it to be land, she then lay down and slept long; when she awoke, her guide the wolf had disappeared, so she conjectured she was on the mainland, and taking her direction from the sun as she would have done in her own country, she proceeded along the coast, till she came to a large River. She then thought to herself "this river comes from the wooded country rising in the mountains there, I will follow it as long as it appears to come from the direction of my country." While walking along she perceived many shining substances, resembling stones. When she sat down to rest, she examined them more closely, and beating them with stones found that they were soft and malleable. So she made collars and other ornaments and hung them round her neck, others she put in her ears like earrings and also round her arms. Then she proceeded on her way. Near there she left the river and proceeded over the barren grounds in the direction she judged lay her people. She determined to mark the way, so that in the event of her finding her people she might be able to find her way back to where this precious substance for making knives and arrowheads was. She consequently built small cairns of loose stones, as landmarks on the tops of the highest hills along her route. After having thus travelled for many days, she at length found a deserted camp, which she knew by certain unmistakable signs to be Yellow Knife, and probably of her own band. The traces were moreover quite fresh. It was no difficult matter for her to follow their track, so on she went with renewed courage. Two days after, on arriving on the top of a high hill, she discovered the never-to-be-forgotten lodges in a cluster down in the valley, and the men, women and boys of all ages, engaged in some athletic games on the adjoining prairie. She immediately advanced and was close, ere they perceived her owing their preoccupation with their games. The Esquimaux! The Esquimaux! was yelled from a hundred throats, but she soon quieted the tumult, by calling to them in their own language her name and that she was alone, and that only her dress was Esquimaux. Her friends then crowded round her, for it proved to be the same band from which she had been kidnapped nearly three years before. Innumerable were the questions put to her of course, regarding her long absence. While told her story the copper ornaments attached to her neck and arms attracted universal admiration. In answer to their questions she told them where she found them and said that the ground along the river was literally covered with it being so plentiful. "Could you find the place again do you think? said they, Oh yes! answered she, and then she described how she had marked her route, with that object. A band of young men were immediately appointed to go and procure a supply of the much coveted metal, and to make assurance doubly sure that they would not lose the way, they proposed to the woman, that she would accompany them. "Yes," said she, I will accompany you, if you will, one and all, promise faithfully to treat, and in every respect regard me, as a man until our return to our people here. The promise was accordingly given, and the party set out. In due time they arrived at the place, and found the river and the copper just as she described it. They soon collected as mush as they could carry, without any apparent diminution of the vast quantities strewed about, and took their way home again. They were all in high good humor with their success, when they encamped the first night on their return. But among a band of men, as among sheep, there are always two or three *black* ones. In the band now in question, there were such, who now, that they had got in their possession that which they were dependent on the woman to find, and now knew the way themselves, thought it time to throw off their reserve and forget their promises. These proposed to the others to amuse themselves with the woman and their scruples were easily, too easily overcome. So they ravished her, and abused her one after the other all night. When morning broke, they prepared to resume their journey, and calling to the woman to come along, they started, but the woman sat still, on the spot they had left her. They thought she was only sulky, and should follow them, but they saw her no more. They returned to their people, and delighted their eyes with the amount of their brilliant spoils. The following year, a party again departed for more Copper, but on their arrival where the woman had been left, they found her seated on the same spot, but quite dead, and sunk about two inches into the mossy ground. On they went and procured another supply at the old place, but it appeared to be less plentiful than before. Year after year they thus went for copper, and year after year they found that the woman had sunk deeper into the ground, and in the same proportion was the difficulty of procuring a supply of the metal till at last only the top of the woman's head was visible, and that year only a few large pieces could be found above ground and these they could neither break, cut nor carry away. With great difficulty they succeeded in getting a little by digging in the ground for it. The following year the woman had disappeared in the ground altogether, and so had the copper, so the party had to return nearly empty handed, for it was too much labor for them to dig in such a soil, to the depth required to find it. After that time they went no more in search of copper but the cairns of stones built on tops of the mountains across the Barren Lands to the Copper Mine River as landmarks to shrew the route, are still standing to this day. So ends the story of the finding and disappearance of the Copper on Coppermine River. It was told to me by two old Yellow Knife woman in Old Cadieu's House one day, and he interpreted for them. You will doubtless think it told in a very prosy style, but I cannot help, I never could make a story short and sweet in writing, in my life without spoiling it. Now you have got it in full, you can cut it down or *Doctor* it any way you choose.

I now send two sheets on the Moose and Carcajou, these are the only two on which I have "said my say", but I have many others begun. I cannot send the sketches now, for I find it impossible to put them in a convenient parcel. They must go in a case in spring.

Three days ago I secured two superb specimens of the Strongwood Carribou. They are now handing frozen in the Store till I find time to try and dry them. You may be sure I will do my best to prepare them well. One is of a young female, the other a young male, going on their third year. I am promised also the skin of a LARGE male, when they kill one, so that you may have the three specimens. In case that Mr. Smith at Fort Rae should not collect in Spring, I have engaged Olivier Laferté, who winters there to do so, also while coming across next spring to collect some geological specimens on that side of the Lake, and also from the Islands in crossing. He is a careful man and I have made it worth his while to give himself a little trouble to please me. I know that Clarke got a great many eggs in those Islands, but were any of them identified? Very few I think. I saw him pack them up here, and he gave names to some of them, that I knew perfectly well, very wide of the mark. From Salt River too I expect to get some prizes, if alls well, but the worst point of those Beaulieus is, that one can never depend upon them.

Dec. 17th I am very sorry at having done so little for yet in the writing way, but I will make up for lost time by and bye. I can assure you that I have been making a regular slave of myself this whole fall, for Mr. King left me all alone, and I had everything to attend to. He has now returned with the Winter Express. He arrived late last night and the packet must leave tomorrow morning. I have received a short note from Macfarlane at Good Hope. He thinks that you or Professor

Baird ought to write to the Govr. about those cases which have been, we hear, lying at Norway House so long. He also thinks it is a damned shame that permission to send us a "horn" to keep the cold out of our toes, has been again refused. He thinks however, that you might "unbeknowns't" like, slip in a little in case without saying anything about it.

In February I am going across to Fort Rae to see Smith, then he will return here with me to meet the Winter Packet. What a pity we won't have a "horn". Then I hope to receive long letters from you.

Reid arrived here with Mr. King and has kept me up the whole blessed night chatting. Thank God he has fallen asleep at last, so I must now seal up and close the packet Box. By the by, Mrs. Reid has another little daughter, three days before he started to come over here. Mrs. Kirby has also brought forth a daughter, the little man is certainly "Ahsteeny eslah".

Wishing you all the compliments of the approaching season, a Merry Christmas and happy New Year. I remain, My Dear old friend Yours' very sincerely,

J. LOCKHART

49. LAWRENCE CLARKE to ROBERT KENNICOTT

Fort a la Corn Jany 16/65

[SIA, HBC Corr Coll, Folder 9]

[Dr. Joel Asaph Allen (1838-1921) was one of Baird's protégés. He became a prominent zoologist, specializing in systematics, and he wrote scores of articles and edited publications such as the American Museum *Bulletin* and *Memoirs*. He was the Curator of Birds at the Harvard Museum of Comparative Zoology from 1867 to 1885, then becoming the Curator of mammalogy and ornithology at the American Museum of Natural History (New York). In this letter Clarke relates his opinion on the effects of Ross' transgressions in the pursuit of natural history specimens. Ross was fined £27.6.3 for misusing company funds while he was in charge of the Mackenzie River District. (H.B.C.A., PAM, B.239/k/3, p. 302).]

My dear Kennycott

Your long deferred but not the less welcome letter came safe to hand. Let me see, this is positively the second letter I have received from you since you departed from amongst us! Why man I begun to have serious doubts of your being in the land of the living, or had not been swallowed up body and bones by some of those terrible southern niggers who are apparently determined to make you northern Gentlemen pay for your whistle, if you succeed in getting it at all. You speak of "*fragments* of broken promises," well I forgive you, tho I must acknowledge to having felt *rather* riled at your long silence.

I give you all the credit in the world my dear fellow for looking after my interests in seeing that I received due credit for my collections, but you know my perfect indifference to seeing my august name in print. I just collected, out of friendship to yourself personally, secondly to oblige Professor Baird, and lastly because I really begun to feel an

Lawrence Clarke Jr. (1832-90) in the 1870s. L. Clarke Collection, Provincial Archives of Manitoba.

interest in the thing myself. As for poor Barney and his superb egotism, I really rather admire than feel angry at it. He is certainly a magnificent impersonification of Self! I am really sorry that that unfortunate Key of fishes should have again gone astray, it together with a small blue case containing a Hare Indians scull was found at the portage after your departure. I took them under my own care as far as Cumberland and from thence sent them on under the special care of L'Esperance to Norway House. I saw the old fellow at the Grand Rapid on my way up from York, after the receipt of your letter, and enquired what he had done with the things, he swears he delivered them to Donald Ross at Norway House. I will make a search in the Norway House stores next summer, for the missing articles. I am as sorry as yourself that you did not take my advice and pass your first summer at Fort Rae, for I believe with you that with the exception of, perhaps, the Anderson, it is the best place in the district for both mammals and Birds, and just conclusion could be formed from the specimens found there of the arctic fauna than any other place in the district. My total ignorance of Natural History was a great bar to my successfully working up the place, had I had another years stay at Rae, or had I had the privilege of being inducted into the pursuit by yourself for some months your first summer, I should have probably exhausted the Natural History of the place.

It is gratifying to me that you found so many undescribed species in the collections I sent out from Rae. I certainly spared neither expense or exertion to gather all and everything I possibly could, knowing it to be my last year in the district.

I heartily congratulate you on becoming the presiding Genus of an Institute of your own starting; it is both alike honorable to the citizens who have shown such confidence in you, and honorable to yourself for having by your energy and ability succeeded in establishing another monument of popular education. Your own collections alone will fill no mean portion of the Museum and will be, not, the least valuable portion of its treasures.

Many thanks for your Kind invitation to the Grove, but I fear circumstances will prevent me for many years, if ever, of availing myself of your kindness.

You mention yourself and Dr. Allen having sent me newspapers, my dear fellow I thank both you and him for your kindness, but some one has cribbed them "en route" for I certainly have never received one. I have written to a couple of my friends in the Saskatchewyan to try to get some "bats", here there is nothing of the kind to be seen, but I beg you will not make any perquisitions on your friend for me, for I may and in this instance, unless my brother officers will fulfil my request, am quite unable to render the assistance asked for, if there was bats here, he should have every mothers son of them I could spiflicate. I shall remember him however the first *bat* I see, then woe betide his mongrel mouship.

I have got a good specimen of the yellowish ground squirrel of Richardsons, it has no spots and I believe the identical gentleman you want.

I am placed in a peculiar position here, and under present circumstances can do very little in working up the Natural History of this place, we are surrounded with fur Traders, are seldom two days together at home and to add to the difficulties my Bourgeois throws cold water on the affair and seems to think, the Goth, that I cannot be a Smithsonian collector and a faithful servant of the Company. The truth is my dear fellow, but this is *entirely entre nous, Barney* has done incalculable damage by his dishonest dealings in "K" it was brot home to him of having misappropriated much of the companys property in obtaining his collections, and has been fined heavily by minutes of council this year; the result is, that people on this side feel an atipathy to meddling with collections of any sort. A few things I have collected, but nothing to what I could or will yet do, if certain changes are made in this district which is in contemplation, remember *all this is strictly private*, but you may hint to Baird some excuses for my seeming laziness.

My better half was quite delighted with your specimens of sempstress work and values it highly. Providence has taken one of our little Twins from us. Your friend Aileen is still a suffern from her sore arm, she with us all often talks and think kindly of you and the good old times.

I am just off a trip of 17 days duration and as soon as the Express passes, which will be on or about the 20th, I am off again for the Lord Knows how long. I have still many letters both official and private to write, so you must excuse this d_____ble screed. I will write you a *ream* in spring.

They are starving like the devil in the Saskatchn. this year, my lucky star is still constant. I have plenty, God be thanked.

McKenzie is at Green Lake and asks me for Gods sake to send him a bag Flour. When he wrote they had only 3 fish in the establishment.

Swanston is fur trading at the pass. I have not heard how he is doing.

We all unite in kind regards. Wife would send you some shoes if she knew how

With every good wish, Yours for *all time*,

L. CLARKE

50. RODERICK MACFARLANE to SPENCER BAIRD

Fort Anderson 8 Feby' 1865
[SIA, RU 7215, Box 14]

My dear Professor Baird,

I again avail myself of the departure of the Lower MacKenzie

Packet, of addressing you very briefly on the subject of the "Anderson Collections". From the subjoined statement, you will perceive that a good many eggs have been traded by me from the Esquimaux, since my return here last September. It is to be regretted, however, that the absence of parent birds, especially of the Geese, will render it a matter of no little difficulty to distinguish the eggs of the different species. A similar difficulty will also arise regarding the Swan Eggs, both species being represented here. I believe that it will be necessary for me to make a journey to the shores of Liverpool Bay myself, ere we can have some points solved at least to my own satisfaction. I cannot however devote next summer to this object, so I intend going again over last summer's ground; but if I am here the following season, I shall certainly go down this river, and explore the Coast and country from which the Esquimaux specimens hitherto received, have been procured. The Season of 1866 will therefore, probably complete the "Anderson Collection", tho' I fear that a good many Birds eggs, etc. will still remain unknown, of those belonging to our Fauna.

I have already packed up 6 Boxes, containing all the birds and small mammals, together with some Marine shells etc. 3 of these are now at Good Hope, while the packet sleds will take the other 3. After they leave on the 10th Inst., I shall begin to pack up the eggs, but a want of cotton for this purpose will occasion some difficulty to find a substitute. I shall however secure them as well as I can, and trust that no serious breakages will take place on their way down to Washington.

I should like the Natural History Society of Montreal, to have some things from my Collections—a Bear, A White and blue fox, one or two specimens of Parry's Marmot, as many hares, together with some eggs of such species as may be wanting in their Collection—but I shall write you more fully on this subject in May next.

One of the Young Eagles about which I wrote you last July, died in August. It was probably killed by the others, as it was found in a partially eaten state, while one of them was seen pecking at and eating the dead carcase. Macleod probably forgot to feed them the previous evening. They are usually fed twice a day on fresh meat; but they consume considerably more food in Summer than in Winter. The old one (from 1863) is a male I think. Last Winter, it passed a very dull winter by itself, and would seldom eat of its own accord, in fact, the meat had to be cut and put into its mouth, and frequently also pressed down its throat. It was also comparatively tame. This Winter however it, as well as the 2 Survivors (a Male and a female) of last Summer's collecting, will eat their food readily enough. A piece of meat is generally thrown to each, which they soon discuss. They also quarrel among themselves when feeding; but agree well on other occasions. They are certainly very vitious at present; the old one, as well as the female referred to, will at once attack with beak and claws any one approaching the bars of their Cage; but in the earlier months of Winter, I could lay hold of any of them, without their offering to trouble me. The female is considerably larger than the males. Perhaps she'll lay some eggs in Spring. I shall put them in their outside cage 2 or 3 weeks hence. They never drink water. I have offered them some on different occasions, but without inducing them to partake of any. They will eat fresh fish with as good a relish as meat, also rabbit and partridges when they can get them. When hungry or annoyed they give utterance to a peculiarly disagreeable note, tho' on other occasions their notes are of a more pleasant kind. I have no doubt they would soon fly away if they had the chance, altho' of course for want of practice, their powers that way are not very strong. I may give you further particulars anent them, if required.

I had almost forgotten to mention that I have secured for your Institution, a mountable specimen of the Polar Bear (*Ursus Maritimus*); it is a male but only of medium size, and was obtained from an Esquimaux, who shot it in Liverpool Bay, in the month of October last. It shall be transmitted to you with the other Collections of 1864-5, by the Company's Boats of next Season. I regret however that it is not of the largest size.

The following list will give you an idea of what has been received since I last addressed you from Fort Good Hope.

Colymbus Admasii? - 2 skins from the Esquimaux
_______ *Arcticus?* - parts of female and about 12 eggs
Somateria V. Nigra - 2 or 3 indifferently prepared skins.
_______ *spectabilis* - 2 skins and 1 egg Esquimaux
Nyctea Nivea - some skins and 1 egg which I however find too small—it is of the same size and color as that of a *Marsh Hawk*
Brachyotus Cassinii? - 1 skin and 3 eggs from Esquimaux
Falco Dawsonii? - 8 eggs, received from Dn. Lowr. Andn.
Lagopus Albus? - some skins and about 100 eggs
Numenius Borealis? - 3 tolerably well prepared skins
_______ *Hudsonicus?* - 3 " Do. and 2 eggs
Limosa Hudsonica? - 2 " Do. - eggs got broken.
Strepsilas Interpres? - 3 : Do. Do. Do.
Bernicla Hutchinsii? - 4 heads and about 80 eggs
_______ *Brenta?* - several Do. Do. 360
Anser Gambellii? - 1 head and Do. 100
Cygnus Buccinator - 7 eggs Received from the Esquimaux
_______ *Americanus* - 12 " Received from the Esquimaux
Plectrophanes Lapponicus - 1 skin and 1 egg Do.
Corvus Carnivorous? - 4 eggs - Lower Anders Esquimx.
Charadrius Viringicus - 4 Do.
Larus [blank]? - about 20 eggs Do. *L. Glaucus?* etc. (2 or 3 species)
Harelda Glacialis? - " 60 " Do. chiefly
Archibuteo [blank]? - 4 eggs (2 nests) Do.

Mergus Americana - 3 supposed eggs from Liverpool Bay. Together with 3 October specimens of the Blue Fox, a young Swan shot last September, and about a dozen Rabbits snared at the Fort in October and November 1864. There are also some marine shells obtained from the Esquimaux, besides some other objects of Natul. History.

I shall now conclude, and trust that I'll be enabled to get everything packed up securely and afterwards forwarded to Fort Good Hope in time for the Boats of 1865. With much esteem, I remain, Very faithfully your's

R. MACFARLANE

51. RODERICK MACFARLANE to SPENCER BAIRD

Fort Anderson 15 May 1865

[SIA, RU 7215, Box 14]

[George Rolleston (1829-81) was a professor of anatomy and physiology at Oxford from 1860 to 1881. He introduced the "type" system into zoological studies. In his system, "representative" species were chosen from each animal order for students to dissect and study. For example, he chose the common pigeon as the representative species of the Order *Aves*.]

My dear Professor Baird,

I received your brief notes of 12/17 Octr. last, a few days ago. The egg drills, blowpipes, labels and Isinglass adhesive plaster therein mentioned, came also to hand at the same time.

The more I become acquainted with the "Anderson Country" and its resources, the firmer grows the conviction, that I shall be obliged to leave it long before anything bordering even on an approximate knowledge of its Natural History, shall be ascertained. In my February letter, I mentioned to you that the summer of 1866 would probably close the "Anderson Collection"; but from letters lately received, it would not at all surprise me, if I were appointed to some other charge next Autumn. I trust not however, as I should much prefer returning to this place.

Herewith, you will find the Packing Account of the Collections made during the past year. They are contained in 20 Cases and Cassettes, which have all been forwarded to and rendered safely at Fort Good Hope, whence they will be only embarked in the Company's Boats for transmission to Red River. I trust that the cases will reach Washington in safety, and that you will be pleased with some, at least, of their contents.

I have this spring also forwarded to Governor Dallas "Hudson's Bay House London", the Barren Ground Bear referred to under No. 2136 of the Collection. I had also secured a Musk Ox and Arctic Reindeer for him; but they were unfortunately destroyed by animals, before I could send for them. I shall however try and oblige him next Season; but as the Governor and Committee in England, have ordered several specimens of this Bear, as well as of the Musk Ox and Reindeer, to be procured for them, I fear that it will be out of my power to obtain any for you this Season, additional to the Barren Ground Bear already promised for the Chicago Museum.

I am much obliged to you for your attention to the requests I made last May in favor of Governor Dallas and Dr. Brewer, and it may also be stated that those requests refer not only to past, but also to any future Collections I may make while stationed in this quarter. I should also like that a good specimen of the Blue and White fox, and a few other mammals, were transmitted to Professor Rolleston of Oxford, England, at your earliest convenience, And I may likewise mention that I have apprized Professor Darey of the Natural History Society of Montreal, of my desire that their Museum should have a "Barren Ground Bear, a Blue and White Fox, a Parry's Marmot, and a couple of Rabbits, together with the skins and eggs of any Arctic birds, comprised in my collections, that might still be wanting in the Museum of the Natural History Society."

You thus see that I have not been backward in availing myself of your kind offer to honor any drafts of this kind, that I might make on the Smithsonian Institution.

In Cassette No. 17 of the Packing Account will be found 5 dozen Reindeer Tongues, 3 dozen of which, I trust, Professor Henry will do me the pleasure to accept, while I regret to say that circumstances prevented me from sending more than 2 dozen for your own acceptance, altho' it was my intention to have sent at least double that number. I trust however that next year will enable me to make up the deficiency.

Many thanks for the *New York Ledger.* I get it as regular as can be expected, in this blessed country, we are always well pleased when 3 out of every 4 papers sent to us reach their destination. The last packet has, however, exceeded the average, 3 numbers only being missing between 11 June and 19 Novr., The oldest and latest date of the batch. I may also remark in passing, that by the Winter packet of this year, I received English letters of 9 November, while as above stated, the 19 November is the date of the last *Ledger.* In any year, you could, if necessary, defer writing to us until the first week in November, and your letters would still be pretty certain of reaching Red River in time for the Northern Packet.

After I had last this pleasure in February, and until the middle of March, the Weather was uncommonly cold—from that time until the 20 April, we had very fine Weather indeed—for the Season and situation. Since then however, it has been at times severely cold, 15° and even 18° below Zero, so that we have not had any thaw to speak of, as yet, this spring. The Season therefore looks as if it would be an unusually late one. Birds are consequently very scarce, in fact, none are to be seen about the fort, except a few Whisky Johns. Crows,

Snowbuntings and some Ptarmigan, these last are now gradually assuming their summer plumage.

I regret to say that I have done little or nothing as yet in the collecting line. I have received 2 Crow's nests, and the female and 4 eggs of the Bird entered as *Falco Dawsonii*? in last Summer's list; but which may really be a Goshawk, tho' it can hardly be an Iceland Falcon. A Whisky John's Nest was found by an Indian about the middle of April; but the eggs (5) unfortunately got broken, while the female Golden Eagle skin packed up in Cassette No. 17, was shot on the upper Anderson, some 40 or 50 miles from here, on the 5 Inst. She was at the time sitting upon her nest, which was built on a very tall pine close to the river bank. The nest is said to have been similar to that described under No. 605 of Collections 1863, and it then contained 1 egg which unfortunately fell and broke as the Indian finder was descending the tree therewith. I have also just learnt that a pair of Eagles of this species, were lately seen occupied in building a nest on a tree in this neighborhood. I shall therefore take care that they are not prematurely driven away, as was the case last year, with a pair that had been similarly employed in building on a tree in this vicinity. We may now therefore assume, with no little degree of confidence, that the *Aquila Canadensis*, at least in this quarter, builds on trees as well as on cliffs, tho' by no means so frequently in the former, as in the last mentioned situation.

Enclosed, I beg to send you a brief paper on the Chipewyan or Montagnais family of Indians, obtained by me from the Revd. Father Petitot of the Roman Catholic mission of Fort Good Hope. The same Gentleman has also promised to let me have some other papers on the Montagnais which, when received, shall be transmitted to the Institution.

The List with Notes of the Collections forwarded this spring, shall be sent to you by post, addressed to the care of Governor MacTavish of Fort Garry, and I trust that the document will reach you in due course.

I may inform you that it is my intention (*D.V.*) to go to Franklin Bay again next June, and I trust that the Expedition will turn out eminently successful. I need not however conceal from you, that I have not the slightest expectation that the General Collections for 1865 will, in quantity and quality, approach those of last season. In fact, if they equal the Collections of 1863, I shall be well satisfied. In any event, I hope that we are destined to make a few new *additions* to the Oology of the "Anderson", in course of the season. But the lateness of the Spring will, I fear, prove unfavorable to our egging operations on the way to, as well as on the shores of, the Arctic Sea.

19 May 1865

The packet for Good Hope starts this evening, I must therefore conclude this scrawl. Since writing the foregoing, it has thawed a good deal. On the 17, a few Canada Geese and Grey Wavies were seen, also 2 ducks, 2 large Gulls and a few Terns. But as there's no open water to be seen anywhere, the Geese were seen to retrace their flight South last night. I shall write more fully in Summer (*D.V.*). I hope you will excuse any errors etc. that may be discovered in the lists etc. And with best wishes, I remain, Very faithfully your's

R. MACFARLANE

52. JAMES LOCKHART to ROBERT KENNICOTT

Fort Resolution 26 June 1865

[SIA, HBC Corr Coll, Folder 26]

[William Thomson Smith was stationed in the Mackenzie River District from 1862 to 1865. In 1865 he was forcibly removed to Red River by his employees at Fort Rae in order to stand trial for the murder of Pierre Gendron, one of the labourers at his post. Company officials subsequently decided that the shooting was accidental. They also determined that the man who pressed charges against Smith, Neil McNevin, did so spitefully. Thomas Hardisty was the apprentice clerk at Fort Rae in 1865-66. (H.B.C.A., PAM, B.239/k/13)]

My Dear Kennicott,

You will know by the date of this letter that we are beginning to look out and prepare for the arrival of the McKenzie's River brigades. As I am not sure about accompanying them to Portage La Loche this summer, I shall write my letter to you at once. Your kind letter of October 24, from Washington reached me by the March packet, accompanied by the ring enclosed in your Tobacco Box for which accept my hearty thanks. The Ring was just the thing I wanted, and will suit the purpose for which I wanted it admirably.

You ask about Fishes, you know that is a dirty business and no one likes to undertake a collection, but altho' nothing has been done as yet, "courage mon amis". Mr. King, the clerk here has promised to begin in earnest, and endeavor to make a full collection, and I have promised to help him. The small ones will be very difficult to preserve having no Alcohol, that sent in some years ago having all leaked out somehow. Jones the rascal disappointed me altogether, for he promised me faithfully to take up that branch. For those who like fish, I am certain that no branch of Nat. His. would be so full of interest as the Fishes of this district. For at every different Fort, the same kinds vary so much in size and shape, that one would almost swear, they were of different species, and doubtless many of them are too.

The longtailed shrew is here, but I have failed to secure a specimen. While I was off goose hunting one was caught here, but the folks being to lazy to skin it laid it aside to await my arrival, but I being detained several days longer than I intended, the shrew got completely spoiled and was thrown away. Better luck next time. I bagged 200 Geese and 3 Swans this spring. I intended to have skinned a lot of them, but they were so fat that they all spoiled on my hands. I have a few yet of the small white wavy. The Fort is overrun with mice. They have completely spoiled five of them for me. I do not pretend to collect here; it is utterly impossible. I have gone out several times, nest hunting, but did not get one. The mosquitoes blinded me, and twice I sunk in the shaking bogs and hardly got out again. I am indeed very angry and disappointed at being able to do nothing, but it is not thro' want of exertion on my part. I have two collecting for me at Fort Rae, perhaps they may get some of the rarities, not yet procured, towards which I particularly directed their attention such as the *Colymbus Adamsii*, threetoed Woodpeckers, crossbills and others.

There was a report current here this winter that I was to be removed to Athabasca District this season, but I have heard nothing about it from the authorities. People are coming here to oppose us from all sides next winter, we hear. The Yankees have found out the road over the mountains to St. John's in Peace River. Last winter several wintered at St. John's, Dunvegan and Vermillion. Next winter they are to cross over to the Liards they say. This morning one of my Indians arrived here from Vermillion, he tells me that there is a large party of Longknives on Peace River, they have many boats and many masters, but one is at the head and conducts all their movements. I wish to goodness that I could get out of the district ere opposition comes in to disturb us.

We were very short of grub this spring since the wildfowl arrived, we have lived entirely on the proceeds of our guns. Now we are getting Trout and a few carp and Inconnu.

Our friends below have all written to you I think, and will give you all the news from that quarter. Here there is no news at all worth mentioning. Everything goes on in the same monotonous way. Our only anxiety being how to procure the wherewith to feed our people. Mrs. Reid had another daughter about Christmas last. Poor Smith, whom you met coming in here, the summer you went out, has met with a melancholy accident. His gun went off in his hands and killed one of his men and some say it was not accidental. The consequence is that poor Smith has been sent out to Red River to stand his trial. Such an occurrence never took place in this district before. It has therefore caused considerable excitement. Some say there was a woman implicated, but I cannot tell you the rights of the case, for I have heard a hundred different versions. I took Smith up to Athabasca in March. I only remained there four days. I returned here from Chipewyan in *six days*.

Did you get the Dogwhip and shoes, I sent you last summer in one of the Smithsonian cases?

I cannot think of anything more to write about at present so I will close with wishing you health and prosperity in all your undertakings, and a good understanding with your Southern brethren. I remain, Ever your's faithfully,

J. LOCKHART

P.S. I forget to tell you that I went across to Fort Rae this last winter to

visit Smith (we had not then heard of the sad events which had taken place there). Nothing would please me, but to go by the Grande Traverse, altho' neither I nor any of my men knew exactly what direction to take to find the Island on the opposite side, none of us ever having been that way. We started at midnight and made good way till sunrise. Then bad weather came on, so that we could not see twenty feet from us. On we went till we got thirsty, then we made a small fire on the ice, having taken a little wood with us, and made a cup of Tea. Towards evening the weather cleared a little, and we found that we were close to the Beaver Lodge which you know is far to the left of the Island we were in search of. We immediately changed our course, but darkness came on, then it blew and snowed and still no sight of land. William McLeod whose dog you used to drive at the Youcon, and Louison Laferté were my companions. After walking till near midnight, I determined to camp so we cleared a place, hauled the dogs round us, placed our sled to windward there without dinner or supper, parched with thirst altho' wet with perspiration, we rolled ourselves in our blankets and went to sleep. What a fearful night that was! Louison and I slept like a pair of Tops, but poor William had to run up and down all night to keep himself from freezing, for he had not changed his socks before lying down. When daylight came, we could just see land in the direction from which we had come and we afterwards found that in the darkness and we had passed within a *[blotted]* of the Island. Three hours walking and running took us to land, where we soon had a large fire, made a large kettle of Tea and ate a large breakfast. Do you remember little Tomiere that Antoine Hoole had at the Youcon. I have got him to replace "Royal" who died last summer, but he is not smart enough for "Prince" and "Dick". My train, however, is the crack one on Great Slave Lake yet. Next winter I have a splendid young dog to put with them. I am always raising good dogs, but the other clerks always beg them from me, and I cannot refuse to give them. Mrs. Hardisty has my old Youcon dogs to drive about to her Rabbit Snares.
Yours Ever,

J. LOCKHART

53. JAMES LOCKHART to SPENCER BAIRD

Fort Resolution, Great Slave Lake July 4, 1865
[SIA, HBC Corr Coll, Folder 26]

My Dear Professor

Your esteemed favor of October came to hand by the March packet and I was glad to see that I had filled another gap in your Series.

I notice your directions in regard to "clean mountable Specimens", and will attend more to that in future; for I find that *egging* is a total failure here.

I receive *Harpers Weekly,* pretty regularly, many thanks for your personal attention to the mailing of them. I may say however, that scarcely one month reaches me complete, generally one or two and sometimes three weeks are wanting. *Yankee Notions* and some other papers that friend Kennicott wrote to me had been sent, I suppose have been appropriated by some unscrupulous individual *en chemin* for I have seen nothing of them.

It was certainly an oversight on my part, not to have sent you any lists of the cases sent from here last summer. There were three cases of last Springs Collection, and three I think, of my former Youcon collections.

Mr. Jones had already informed me of his improvement on the vent of the Cartridge cases of the Maynard Rifle. I immediately acted on his suggestion, and now it astonishes the natives not a little.

I am altogether discouraged with the results of my collecting here. I had two men engaged to collect at Fort Rae and one at Salt River. Owing to the lateness of the Spring however, nothing was done. I am sorry to say that I cannot announce one single rarity this year, but I think you will find some good mountable specimens among them. I am sending 4 cases from here. I enclose a List of their general contents. The two Carriboo, I prepared as well as I could. They are both of the same age (going on the third year). I did succeed in getting a full grown one as I intended.

The mice unfortunately found an entrance into Mr. J. Flett's case this winter and damaged some of the specimens. I had them put into a good case however, ere they do much harm.

Mr. Kennicott writes me that the greatest want felt in your scientific world now, is the Northern Fishes. I must confess, that we have all shown extreme reluctance to undertake that branch. Jones promised me four years ago that he would take it up, but has never done. The difficulties attending their collection and preservation have hitherto deterred many of us from entering on it. The alcohol sent us some years ago, you are no doubt aware, all *leaked out* long ago, and "the powers that *were*" put their *veto* on the supply being renewed. Perhaps those now at the head of our affairs might, if applied to, be less strict in that respect. In the meantime, the clerk here, Mr. King, and I have determined to make a collection of the skins of all the Larger Fish in this vicinity. And we will endeavor to induce our friends at other Forts to aid us. So next Spring you may expect a case of two of them. But at best, they cannot but be most unsatisfactory, owing to the scarcity of alum, saltpetre and other preservatives.

I have just this moment received the large red billed Tern *Sterna Caspia*, and Eggs. I fear however, that I will not get them skinned in time to go out this spring for we are expecting the boats momentarily. They are already a week beyond the usual time. As you have already got them however, it will not matter much.

We have had a very late spring this year. The ice is still in view on the Lake. The larger kinds of wildfowl were numerous here this season. My own hunt consisted of 200 geese, 3 Swans, 2 Cranes and ducks innumerable. All the little white wavies that I killed, I skinned, be they bad or good. The mice destroyed six skins for me unfortunately.

I enclose a detailed List of what I have collected this year. Trusting they may all reach you safely and in order. I remain, Your's very truly

J. LOCKHART

54. STRACHAN JONES to SPENCER BAIRD

Fort Rae 27th Nov. 1865

[SIA, HBC Corr Coll, Folder 24]

[Colonel Charles L. Bulkley was the engineer-in-chief of the expedition sent out by the Western Union Telegraph Extension Company to build a telegraph line from the United States, through British Columbia and Russian America, to Siberia and eventually Europe. Kennicott was placed in charge of the scientific field corps attached to the W.U.T.C. expedition. (Neering, *Continental Dash*, 1989, p. 27)]

My dear Professor Baird

You will perceive from my address that I have left the Youcon, but as far as the Smithsonian Institution is concerned it will make no difference for my successor Mr. Jas. McDougal is a most *determined* Collector. He is a Scotsman and a splendid climber among the rocks and fairly astonished the Indians, altho there were great hunters among them, by the way he went up and down the rocks. I was down the Youcon and made a tolerable collection myself, and an Indian that I had sent to the mountains to collect made a first rate one. I had not time before I left the Youcon to examine it properly, but I saw several of the Mountain Plover and their eggs done. Rock Partridge and eggs, and many kinds of hawks, some that I did not know, my own collection is principally of hawks and owls and a few ducks. I got two nests of the small hawk owl with both Parents, a large owl and eggs, a black gull and eggs, three Husky geese, a large Canada goose, a large fossil Beaver thigh bone about a foot long, and a lot of hawks, gulls and ducks eggs and parents. I should not omit to mention that I got a very small owl, one of the Hawk Owls, it is not more than half as big as the other kinds of hawk owl. Perhaps it may be another species for the marks and spots are not quite the same in number, but still it may be a bird not yet fully fledged tho it was fuller in the spring when the other owls were nesting. Mr. McDougal has a good collection too, and together with mine will make as large and as valuable a one as ever was sent from the Youcon. My initials are not on all of my eggs, but Mr. McDougal promises to do that for me and send the whole out by the steamer. I forgot to say that I have a bottle full of small shrews, about a dozen of them gathered at the Fort. They are with the rest at the Youcon. I was very sorry to leave without seeing Mr. Kennicott, but I had insisted upon being removed and then I could not insist upon *remaining* afterwards. I was down the Youcon River a long way and left guides to bring up the steamer, also Indians to hunt and furnish them provisions in passing, and I distributed the Photographs of Captain Bulkly all through the Indians close to the Russian Fort.

I wrote the paper on the Louchew, but sent it by the packet. I could not send the Traineau, canoe, dresses etc. by the same way, but I will send them by the next Portage boats. I hope it was found sufficiently elaborate. If not I shall be happy to afford Mr. Gibb any further information he may require, if he will let me know on what particular subject he requires it. It is very difficult for one, who like myself has spent many years among the Indians, to write a particular account, for the reason that many things we have become so accustomed to as scarcely to remark. These would probably be the first to strike a stranger. I can only say for my paper on the Louchew that I will vouch for its correctness as far as it goes. I shall of course collect all I can for the Smithsn. Institn. as long as I remain in the North.

The double barrel guns came in at last. I sent the one addressed to me down to the Youcon to trade eggs with, as I thought it was of more value there than here.

I am exceedingly obliged to you for the trouble you have taken in procuring a short barrel for my rifle. I have long wished to have one. It would be even more useful than the rifle barrel. As soon as I know the price I shall send you a draft. I should do so at once, but I left the pamphlet with all the prices at the Youcon.

I think Captain Bulkly and Mr. Kennicott will find it very difficult to bring up a Steamer, even as far as Fort Youcon, on account of a rapid near the sea. I have not been quite to that rapid, but not very far off. There is another rapid higher up. I have been about three hundred miles below that rapid, it is not a very bad one. This last mentioned rapid is on the Youcon, and about 80 miles below where the Gens de Butte River joins the Youcon, both rivers are very broad, and when they join you would fancy you were upon a lake if it were not for the strong current. There are many large rivers fall into it all along, to within two short days from the Russian fort where I turned. I might tell you the Indian names of some of these rivers, but it would be of little use as Captain Bulkly has before this given them all english names. I think it is quite practicable, the making a Telegraph line up the Youcon. The greatest difficulty in my opinion would be to guard against *fire* for the Indians often set the woods on fire, and to lay the wire under the surface of the ground would be to ensure its being dug up and made into collar wire and bows for kettles, arrowheads etc. by the Indians. I cannot suggest a remedy. I leave that to the well known ingenuity of the Americans to discover.

I am very happy to hear that the Chicago Museum has been a successful undertaking. Indeed I was quite satisfied it would be so from the time I first heard of it, from my knowledge of Mr. Kennicotts energy and ability; I assure you he made some of us up here *look alive* (to use a common expression), and we do not at all consider ourselves *slow coaches* either. I hope he has taken a good supply of provision with him if he is to winter at Francis Lake. That place has not got a very good name for provisions. They say Fort Rae is a good place for eggs. It may be so; one thing I am certain of, that is, it is a great place for *mice*, for the fort is all alive with them.

Hoping my last collection may prove valuable and arrive safe.

I remain, Yours most sincerely,

STRACHAN JONES

55. DONALD GUNN to SPENCER BAIRD

Red River January 3rd 1866
[SIA, HCB Corr Coll, Folder 20]

My Dear Prof

Your note of the 12 Nov. last in which you acknowledge the receipt of mine of the 15th Septr. came to hand a few days ago. I am sorry that the case was not forwarded from Fort Garry as several opportunities of passing it presented themselves during the latter part of Septr. and Octr. But, I fear, Chief Factor Clare does not interest himself much in our affairs. I think, had Govr Mactavish been in Charge as formerly, that the Box with those which are at Fort Garry from McKenzie's River would have been sent on. I saw Mr. Clare a few days ago. He promised to send the case by the first chance that will offer in the Spring. Stating at the same time, that I must be responsible for the freight—to which I agreed.

I will now endeavour to gratify, so far, as I can your laudable curiosity. From the brevity of my notes, I cannot enter into anything worthy of being dignified with the name of description. I cannot say anything about no. 22 further than that I think it may be a *Fulix* having secured the head and wing. I did not take any description being satisfied that when you see these you would identify the birds. I often feel the want of a short treatise on ornithology, being altogether unable to identify many birds which fall into My hands. No. 20 is the *Spatula Clypeata.* I have a damaged male skin, in plumage it agrees, but in measures differs, from those given in the U.S.P.R.R. *[United States Pacific Railroad Reports]* Exp—dimensions there given being 20 wing 9.50. My Specimen measured thus—Female 16 In, wing 9 In, Male 16 1/2 wing 9 1/2—bill broadest at the point. Lamallae much developed. No. 32 Must be the *Erismatura Rubida* or Ruddy Duck. Female 15 3/4 In., Extent of Wings 20 1/2 In. Its Eggs are large round and white, as many as 20 are sometimes found in one Nest. It is rather rare in our Rivers and Lakes. We found two Nests, one with two Eggs in Shoal Lake another with five in Manitowaba Lake, each Nest was made on the top of a deserted Muskrat house. No. 35 is not the Sheldrake or Fish duck. It is either the red-head or Canvas back. The Indians class it with the fall Ducks, *Fuligulinae*, tho' the Eggs sent were found in two nests, one in Shoal Lake, the other containing 20 Eggs in Manitowaba Lake. It makes its nest on Old Rat houses. Having no specimen before me to refer to, I cannot say what the large Grebe is in size, it stands between *Podiceps occidentalis* and *Podiceps Clarke*; two of my specimens measure, the one 25 x 33, the other 25 1/2 x 33 1/2, without Crest so far as I recollect. These large Grebes are not seen any where in this region except at Shoal Lake and on Devil's Lake in Dacotah Territory. The red necked Grebe is a much smaller bird and is found on many of the ponds and creeks in this region. I have sent the head, neck and wing of a small crested Grebe. It is either the horned or crested Grebe. All these grebes make their nests on tufts of Bulrushes floating in still water and rising as the water rises and falling as it falls. No. 3.4.5.6. are godwits. We shot some of them, two females, on the 6th of June, from appearance they had not then commenced laying. The other two we shot about the 20th, but we could not find any of their nests. It makes it's abode in the barest and most exposed places and lakes thus alarmed long before the traveller can approach near to its nest. We will endeavour to find its Eggs. I know and lament that, there are so many birds and their nests in this region, that have not yet been brought to light, especially the small birds that make their nests here, both in the plains and in woods. I have often endeavoured to interest some of My acquaintances in collecting birds and Eggs, but without success, except in so far as some Members of My own family have lent their aid. I hope the specimens of last summer will get to you safe. The water in Shoal Lake is impregnated with some Saline matter. Having to drink it, was a great draw-back on my enjoyment while there, but at Lake Manitowaba the water was good and we passed the time pleasantly among the Gulls.

At present, I cannot foresee anything that may prevent or forbid undertaking another trip next summer, via Shoal Lake, to Reed Lake. Likely an appropriation to the amt of that of last year will be sufficient. The road from Shoal Lake to Reed Lake is thro woods and rarely travelled with carts. There may be some clearing to do, if so, time will be required and Men must be paid for their time and labor. Reed Lake is not put down on Hind's map. It lies between the N.E. end of Shoal Lake and Lake Winnipeg, between the sources of Nettly Creek and burntwood Lake. If I go on the intended expedition, next summer, my stock of cotton, for stuffing, will not be sufficient. You may likely be able to get some friend at St. Pauls to send over a few pounds early in May, if not I must try Govr Mactavish for some. I have been looking out the whole fall for any stray young terns that might be hovering over the River, but I had no success. A chance Caspian Tern or Sandwich tern is seen occasionally fly over the River. Hitherto I have not been able to secure any of them. The *Scotite American* comes to hand pretty regularly.

Please let me know as soon as convenient what is to be done next summer, and what is to be sought most after. Did you ever find the Eggs of the Mourning Warbler?

Wishing you and your's a happy new year. I Remain, My Dear Prof,
Yours Sincerly

D. GUNN

56. RODERICK MACFARLANE to SPENCER BAIRD

Fort Anderson 20 January 1866
[SIA, RU 7215, Box 14]

My dear Professor Baird,

I write now merely to inform you that I have quite recovered, thank God, from the attack of Fever and Influenza, under which I suffered when I addressed you from Fort Good Hope, in the month of September last.

Influenza, Scarlatina and Measles have since then universally prevailed throughout this extensive District; and have carried off great numbers of Indians and Esquimaux. The Season 1865 has probably been the most calamitous, in this respect, ever experienced by the natives of the MacKenzie River District. At this place we have heard of the death of 55 Indians and Esquimaux; but as we have not yet seen any of the Coast tribes, I dare say that many of them have likewise perished. Unfortunately for us, these maladies broke out and continued here, during the best period of the Season for hunting Reindeer, and trapping Foxes—so that no one was capable of moving about. We are consequently but badly off for Provisions, while our fur prospects are far from bright. I trust however that matters will improve by and bye. I am sorry to say that our principal Chief died here on 6 November (I returned here from Good Hope on the 1st of that month). He was a fine man, and decidedly the *best* "Collector" of "Specimens", among all the Esquimaux.

The very few Esquimaux who were able to visit us in the earlier part of Winter, brought in lots of eggs of *Bernicla Canadensis, B. Hutchinsii, B. Nigricans, Anser Gambellii*, and I think also *A. Hyperboreus*. For want of voyaging provisions however, I much fear that it will be impossible for us, to render the whole of "Collection 1865" at Fort Good Hope, betwixt and spring. I shall do what I can; but despite my best efforts, I am afraid that several cases will have to remain inland until next year.

In my fall letter, I mentioned that I thought you might reckon on my going to Churchill next year. I have not however yet *quite* decided on doing so. My health is not now what it once was, and there are other things also to be thot of in connection therewith; but as I shall (*D.V.*) be going to the Portage this summer, where I will probably meet Governor Mactavish, who is expected on a visit North. I will thence address you further on this subject. Mr. Lockhart expects to obtain "leave of absence" for next "Outfit", and Mr. Thomas Hardisty will also be leaving for Canada, by the Portage Brigades.

We have neither seen nor heard anything of Mr. Kennicott. I trust that I shall hear from him when the Youcon Packet arrives. I can now scarcely expect to have the pleasure of meeting him unless indeed, he comes up along with the packet people, which is not at all likely.

Mr. Hardisty, Mrs. Gaudet and several others, very narrowly escaped being carried off by Scarlatina last Autumn. The servants however lost many of their children.

Trusting to have the pleasure of writing to you at great length, on the Subject of Collection etc. next Summer. I now beg to subscribe myself, With much esteem, Most faithfully your's

R. MACFARLANE

57. JAMES LOCKHART to SPENCER BAIRD

Fort Resolution Great Slave Lake 18 December 1866

[SIA, HBC Corr Coll, Folder 26]

[Lockhart refers to an aborted meeting with Kennicott in this letter. What he did not know was that Kennicott failed to arrive at Fort Yukon in July because he had died near Nulato on 13 May.]

My Dear Professor,

I have to acknowledge receipt of your esteemed favor of 11 April, with enclosures of Drills and blowpipes and accompanying List of Birds, also of Kennicott's note from "off Graytown, San Francisco". All these gave me much pleasure. I was however, very sorry to learn that my Youcon collection has suffered so much from breakage. You would not have been surprised I think at that result, could you have witnessed the thumping they unavoidably received, crossing the numerous portages from the Youcon to Red River, indeed I might say to Washington.

I note your remarks anent the various desiderata in the specimen line. With regard to the little goose you mention, I cannot speak with any certainty, for there seems to me to be an *infinite* variety of the Gray Geese, differing in size, shape and color. Last spring we killed several of a very small species, little larger than a mallard. All these had more or less white along the shoulder or front edge of the wing. Should I fall in with any of these next Spring, I will not fail to send some for your inspection. Last spring also for the first time, I saw a black *archebuteo* while out hunting with two Indians. It flew up quite near us, but I was unprepared. The Indians both fired, but missed. They both told me that they had never seen any of the same kind before.

Many thanks for your thoughtful kindness in ordering a shot barrel for my rifle. I have long wished for one, but felt diffident to trouble you about it.

Some of the long looked for cases from the Smithsonian Institution came in this last summer, but Mr. Hardisty will doubtless inform you how the contents were distributed, for I was not present. I think however that *all* the trading goods were handed over to McFarlane to aid in paying off some of his outstanding debts for Eggs etc.

We have had a dreadful time of it this last summer and fall. Scarlet Fever and Measles have reached this district at last, and have caused a great mortality among the Indians. Strange to say, I was the first, in this district to be seized with the Fever. A day or two after I last wrote to you, and while still in expectation of the arrival of the McKenzie River brigades on their way to portage La Loche, I suddenly fell ill, and from our having neither Medicine nor medical advice and utter ignorance of the nature of the malady, I suffered long and severely. Nor have I yet quite recovered from the effects of it. Hence my failure now to send you a continuance of the "Notes" which you were pleased to approve of. Mr. Hardisty fell ill immediately on his return from the portage; by letters received today I learn that he has recovered, but Messrs. Macfarlane, McLean and Brass are all ill of the same disease. Very few Europeans or Whites have died of it, but the poor Indians succumb at once.

Mr. Jones came up from the Youcon last fall and is now wintering at Fort Rae. He remained at Youcon till 10 July (the date Kennicott wrote that he expected to be there), in expectation of his arrival, but our friend did not make his appearance. Jones could defer his departure no longer, or he would have missed the Peel's River Boats. I suspect that Kennicott had much more difficulty in surmounting the rapids, falls and other obstacles on his way up the Youcon, than he anticipated. Jones did all he could however, by sending guides and hunters down the River to meet the party, ere he left. I trust they may succeed to their hearts content.

I hope in March to hear of the safe arrival at the Smithsonian, of the cases sent out last summer.

I get furlough next summer, so you must not be surprised, if, some time next winter the Northern Bear should drop in to give you a paw.

Wishing you all the compliments of the approaching festive Season.

I remain, Yours most Sincerely,

J. LOCKHART

58. WILLIAM MACTAVISH to SPENCER BAIRD

Fort Garry 11 Feby 1867

[SIA, HBC Corr. Coll., Folder 33]

[William Mctavish married Mary Sally McDermot in 1864. She was the daughter of Andrew McDermot, a prominent Red River merchant. The Mr. Scoville referred to in this letter may have been Jonathan Scoville (1830-91), a U.S. congressional representative and manufacturer from New York. The Collins Telegraph Survey that Scoville was supposedly associated with was the scheme devised by Perry McDonough Collins, and sponsored by the Western Union Telegraph Extension Company, in order to determine if communications could be established between North America and Europe via Russian America, Berings Strait and Siberia.]

My Dear Professor Baird

Since writing you last I have been determined to write again on a point or two which I had then omitted to touch on. On one of them I was admit wilfully silent, the others I forgot or rather overlooked. The former was your questions regarding sending Whiskey to Hardisty. In my position I thought it best to let things take their course in silence. But as this is a sneaking *[faded]* way to get out of it, I will say that it is contrary to rule to send spirits of any kind into McKenzie River except for medicinal purposes. So that if as a medical man you consider Hardistys ailments require something of the kind, I may tell you that packages for the Companys officers are never subjected to examination by us. Hardisty is an Officer in the service and knows his duty as well as I do. Another point I omitted was to beg you to send, for me, by mail, from the patent Office, a few garden seeds say onions, carrots, cauliflower, cucumber, melon and Tomatoes, with a few hardy annual flowers seeds. Of course if they are so to be had I would willingly, and if you will allow me to say it, prefer to pay for them. The other item I omitted is of a more delicate nature. My wife for some years past been occasionally liable to attacks of illness which appears to be something like Hysteria. She has occasionally the peculiar fit, followed by longer or shorter intervals of unconsciousness, and as I am getting anxious on her account, and the medical men here seem unable to do anything for her, I would like to try the effect of change and to consult some authority on this peculiar affection. She is so nervous that I would not like to take her across the Atlantic. I have thought that possibly you might be able to inform me of some medical man in the States who has a reputation for skill in this peculiar affection, or one in whose opinion I might place implied confidence. Could you do so and add another to my many obligations to you?

Since last writing you I have learned that a person calling himself Scoville has been forwarded through the country by the Hudsons Bay Company people at the Companys expense. He gave himself out to be connected with the Collins Telegraph Survey preceding to Washington on duty. At each Post he got supplies and was furnished with men horses, dogs or whatever he required to carry him on. The charge for all of which was given him the shape of an amount from each Post, all of which he was to present to me and settle with me for the amount when he came to the settlement. He never came near me or hinted in any way that he had been assisted by the Company, and I am as yet at a loss to find out what he did with the horses he got at the last Post (Fort Ellice) to bring him here. On his way down above Fort Ellice he had, with the Companys people sent with him, gone to one of the Horse guards and taken from there two superior horses reserved for a special purpose, leaving in their stead two of those furnished him at the Qu'appele Post, and I am convinced that Mr. Scoville was an adventurer who has practised on the credulity of the Companys people, but mention the circumstance to you as you may be able to inform me if there was such a person attached to the Telegraph Survey. I am as yet imperfectly informed on the matter, and I am at a loss to see how he contrived to make his first start at the Companys expense. Once started by the Company I can see how arriving at a Post with the Companys servants specially sent he would have little difficulty in being sent on in the same manner. This is the second instance of the kind, but the first, which occurred last year, was not carried to anything like the extent this has been. It is very provoking and will lead to pretty stringent regulations on the subject, indeed to a positive order to give no assistance to any strangers.

I am very much obliged to you for the Smithsonian *Report* 1866 which reached me by the last Mail, and I herewith return to you the printed form executed which I received at the same time under cover to my address.

I am having rather a busy time of it from agitation here, the reflex of the debates in the Canadian Commons on this country. They may spout as they please, but whether this country is at once admitted into the Confederation or made into a crown colony its ultimate resting place will be under the stars and stripes, so you may look at the specimens from it as belonging to the fauna of your Country in the end. I see however you are willing to buy it, or at all events, it is proposed that you should be willing, which the Canadians are not, and if they were willing, are unable to pay. But if said thus I fancy Loyal subjects of the Queen would have me hung for a Traitor. So I must struggle on heartily sick of it, my notions of honor in man becoming hourly more unsettled.

We have had a very severe winter so far and a great deal of snow. The mean of the Ther. for January by my own observations was -35.3° the lowest was -43° on the 12 Jany. On the night of the 8 February it was -43.5°, tolerably sharp for Red River, though in some parts of the country it would be considered moderate. With Kindest regards,

Believe me, Yours Faithfully

W. MACTAVISH

59. DONALD GUNN to SPENCER BAIRD

Red River March 1st 1867

[SIA, HBC Corr Coll, Folder 20]

[Observations on native interment rituals included in this letter were incorporated into an article written by Gunn entitled "Indian remains near Red River settlement, Hudson's Bay Territory," *SIAR*, 1867, pp. 399-400.]

My Dear Prof.

Your two letters the one of Novr 3rd the other of Novr 24th enclosing the splendid case of instruments which you so kindly sent to me—I can assure you the case and its contents are far superior to any that I expected, and permit me to add, superior to anything that I deserved, and will ever be to me a proof of your kindness and liberality—and for which in the mean time I can only entreat of you to accept my sincere and unfeigned thanks, and if I live until old age lays me on the shelf when I can skin neither bird nor beast, I will shew the case with it's contents to my friends and say with pride they are Prof Baird's gifts.

I cannot describe my feelings on reading the sad tidings which your letters conveyed. It being the first intimation which I had of the Melancholly event. Science has, in him, lost one of its ablest and most indefatigable students, and all who had the felicity of being intimately acquainted with him, lost an amiable and warm hearted friend. Poor Kennicott little did I think, when I saw him on his return from the North, that we were doomed never to see each other again in this life. He was then proposing to return in a year or two, and proceed to explore the shores of Hudson's Bay. I deeply sympathise with his bereaved relatives; the news of his fate must have been, to them, a severe and sad blow indeed.

I have been collecting a few Indian relicts of former ages (ie) stone axes, Mallets, and sculls. The axes are made of a fine ground blue stone, the mallets of gneise, the sculls were taken from what are apparently sepulchoul mounds. Last October, a neighbour, living on the East side of the River, found that he required an additional cellar to enable him to preserve his root crop from the Winter frosts, and for the accomplishment of his object he commenced digging on the top of a knowl in the woods Close to his field, taking out 8 feet on the square. He did not digg altogether in the Centre, but some-what to one side, on digging down he was rather puzled by the depth of the surface soil, or black vegatable mould, being so much deeper there than he ever found it to be any where else. However, he continued digging down and pitching out, until he got from 4 to 5 ft deep, there he began to cut through decayed wood, apparantley oak; which had been laid in a horizontal position, and on getting a foot or so below the decayed wood in passing down the side of the pit he came on a human scull, having its lower jaw attached, and lower down the vertabrae, Showing that the dead had been placed in a sitting posture. In digging down he found other human remains and at a depth of 8 feet from the top of the tumulus and on a level with the surface of the surrounding country he struck on a floor of very smooth and hard white mud which, looked as if hardened by the action of fire, for bits of coal were found on it. On this clay flooring 4 or 5 sculls were found lying on the face, with small bones, those of the fingers and toes, an Earthen Kettle with a shell in it, such as live at present in this River. Beaver bones, 2 stone pipes (of fine blue stone without a heel) perforated through, three ornaments made of shell or bone, two of them I think are of shell, likely the shells of the small Turtles that are found in the River here, the other must be bone and is about 5 In in length, one perforated shell used for ornament, a few Beads made of shell. There is another Tumulus four or five hundred yards from the former, directly south of it. It is larger than the one that has been opened, and I believe, that if opened something interesting will be found in it. These mounds were known for many years past, but never supposed to have been a work of art, or raised by human labour, but now I begin to entertain the opinion that many such sepulchial mounds are to be found in this vicinity.

The Indians dwelling in this section of the country have no traditional knowledge relating to these mounds, when any questions are put to them, as to the time when erected and the use for which they were raised. They answer that they were mud dwellings such as are occupied, at present, by the Mandans on the upper Missouri, and that they had been built very long ago. Who the builders were they know not. This ignorance of former times and transactions can to some extent be pretty satisfactorily accounted for from the well known fact that this region has often had a change of inhabitants since the advent of the whites. The Crees were in possession when the first Traders found their way to Lake Winnipeg as they then called it. The Assinniboines succeeded the Crees, the latter tribes breaking off or separating from their kindred the dakotas. The Sauteux or Ojibois followed in wake of the traders from Canada, chiefly in the last decades of the last century. If ever the Cree or the Assinniboines deposited their dead under mounds of earth, They discontinued that mode of interrment long before the advent of the whites, otherwise both Missionaries and Traders must have seen and recorded the customs, or even recent traces of it. In absence of all testimony we are led to the conclusion that, they are monuments of considerable antiquity, and, that, the race who raised them and whose remains they cover have passed away, ages since, or became mixed up with a race or races, if not more barbarous, evidently less energetic and industrious, who did not manifest their affectionate regard for the dead, by performing so much labor in covering their remains. May we not, with some reason, conjecture: that the object for which these mounds were *heaped* up, with so much toil, was to transmit to generations, then unborn, the fame of some renowned patriot warior who led the wariors of his tribes to combat against encroaching foes and who was victorious in the strife. The first thought that occurs to one's mind in examining these tokens of mortality is that, they were erected, or in course of Erection during a long period of time—that successive generations took each a share of the work—burried their dead over those who had been deposited before them, and added their share of the Earth until the practice of that mode of interment had fallen into disuse. Yet on reflecting over the subject I am disposed to come to the conclusion that these mounds were raised over the remains of men who stood high in the esteem of their family, and tribe—who laboured to build them tombs from the laudable motive of perpetuating the memory of friends and benefactors. The tombs might have been reopened to receive the remains of the family or kindred of the first occupier, and to whose memory the tomb had been erected—or at periods of time much more recent. The Indians might prefer entering in these mounds finding them dry and easier to dig than the surrounding soil. Such seems to have been the case in regard to the great Tumulus on the west side of this River sometime about 1786 the small pox spread over what is now known as Hudson Bay Territory carrying off the natives by thousands. the Crees at that time occupied this vicinity. I have seen and spoken to an old man, the only surviver out of many tents. He stated that at the commencement of the mortality the Indians for sometime burried in the above mounds, but did not erect the same, and that at a latter period of the disease the living did not attempt burrying the dead.

D. GUNN

[Addendum]
Up to the last years of the last century the Indians on the western Shores of Hudson's Bay occasionally disposed of their dead by placing them on scaffolds or stages. I am not aware that they ever returned to look after the bones for the purpose of interring them. The Indians occupying this part of the Country at present, Inter the dead, but never, to my knowledge, in a sitting posture. They press the knees up towards the chin, in a word, they roll up the dead into the smallest possible space open a shallow grave, lay the body to it's side with the face as a general thing towards the East. The Assinniboines still continue the custom of laying their dead on stages. I have been carried away by this subject to an unreasonable length, knowing the value of time to you. I ought to have said an unexcusable length, and for this sin of commission I entreat your kind indulgence. I long much for the accomplishment of the intended expedition but; Anticipate the pleasure of hearing from you some time in May before I leave for the breeding grounds and trust that you will let me know if you are, or will be, desirous of having any of these antiquites of which I have written, viz Sculls, pipes, ornaments, etc. If you desire it I will search into some other Mounds in this vicinity and send whatever I find, that may appear to me worth the cost of Transportation. The winter has been late in setting in we had very little snow before the beginning of January. The snow is very little in the Settlement, but is spoken of as being very deep towards the Lake of the Woods and in the plains towards the Missisouri. The cold has not been extremely intense but very regular; this is the 22nd of March, and we have not had so much thaw as to wet the soles of our Mocasins.

With very best Wishes I Remain, My Dear Prof, Yours Most Sincerely

D. GUNN

60. DONALD GUNN to SPENCER BAIRD

Red River Septr 26th/68
[SIA, HBC Corr Coll, Folder 20]
[The "Narration" mentioned by Gunn in this letter refers to his article entitled "An egging exploration to Shoal Lake," *SIAR*, 1867, pp. 427-32. Dr. William Cowan (1818-1902) went to Rupert's Land with a group of military pensioners in 1849. He joined the HBC during the 1850s, becoming a Chief Trader in 1860. From 1862 to 1870 he was posted to Fort Garry where he reported directly to Governor Mactavish. He was one of the founding members of the Historical and Scientific Society of Manitoba, when it was organized formally in 1879. Samuel H. Scudder of Boston was one of Baird's entomological collaborators. He compiled a *Catalogue of the Orthoptera of North America* (1868), as well as *Nomenclator Zoologicus* (1882) which contained approximately 80,000 names of living and fossil animals.]

My Dear Prof

I believe that I am your debtor for two letters, to neither of which, I fear, that I have not returned any answer yet. Your note of May acknowledging the receipt of My Narration of Trip to the Lakes came to hand in the usual time. But I had nothing of any Interest to communicate and for that reason did not write, believing that, some future time would serve as well to tender you my thanks for your kindness in promising to "emend and curtail" my very imperfect sketch before you present it to the eye of the learned public, for I was very sensible of it's many imperfections, and truly glad to learn that you were so kindly disposed towards Me; as to take the the trouble of revising it and throwing it into a more readable state. Some time in July 67 I handed the two boxes of Specimens to the Captain of the steamer, as it was on it's way (as we believed) then, to George Town on the upper Red River, but from some reason or other, it was sent back from upper Fort Garry and to my great disappointment I saw it pass down the following day. I had some hope that Dr. Cowan, who is in charge of affairs in R.R., would find an opportunity of forwarding the boxes by some one of the Many trains that went to St. Cloud, in the fall of that year, but being at upper Fort Garry in the first part of last June, I learned from Governor Mactavish that the Boxes were there still. But he kindly promised to forward them by the Steamer on her return trip to George Town. And I was very happy to learn from your note of the 14th July that he made his promise good; and that, the specimens were in your possession and in some degree to your satisfaction.

About the 20th Ulto I forwarded a small case containing the Canada Jay's Nest and Eggs, 3 Eggs of the Eared owl, the Embryoes taken out the Eggs are also sent in a vial, the Owl had it's nest in the fork of a tall poplar tree. There are some Carolina Rail Eggs, likewise two vials containg Grasshoppers and Grasshopper Eggs. These Eggs, I think, must have the embryoes in various stages of their growth. I labelled the vials as to Mr. Scuder, for he cannot fail taking the greatest possible interest in examining these Eggs and their contents. There are some Beetles in the small vial. In the other vial there are full grown Grasshoppers, two or three of a specie altogether unknown to us. They appeared in June. They are larger than the common Grasshopper, color between blue and brown; under wings purple, with another creature apparently of the Locust family it is green all over and was never observed in this locality before this season. The case I put into the hands of a trusty fellow going directly to St. Cloud, and told him to hand it in to one of the Express offices. And that, I believe, he will do, and you may before now have the treasure in your own possession. I have, this summer, devoted as much of My time and attention to the habits and development of the Grass-hoppers as possible, yet all my additional knowledge is very triffling. The Creature on its emancipation from the Egg or Nest takes one, two or three active leaps, and then rests for a few minutes, as if over powered by the effort it had already made. It's color is pale white like the Egg. I think that exposure to the sun and air for a few days is the cause of its assuming a brown color. I am almost certain that it Moults only once and it does so when its Wings are full grown. The Creature attaches itself to a willow or any plant near it, and there works itself out of its Case which it leaves Attached to the plant, it continues for a few days very weak and Slugish and is easily known by the bright scarlet spot on the head and between the heads of the wings. The wings take a week or 10 days to dry, so as to enable the creature to take its flight, to look out for some unexhausted region where to deposit it's eggs. I have reason to believe that they never deposite their Eggs year after year in the same District of the Country. Permit me further to observe that all the Eggs in a Nest do not become enduced with life the same day, nor perhaps the same week. Those in the bottom of the Nest remaining in the shell while those who were in upper part of it are for some time enjoying light and air and all the other delicacies of the opening Summer. As soon as they began to move in the spring they travelled regularly to the South East, plunging boldly into the River. Few, if any of them, gained the Eastern bank. Thousands of them were devoured by fish. Myriads of them were drowned and cast back on the shores where they lay putrifying in heaps and poisoning the air in every direction. They take from fifty to fifty five days to get their wings. And when the wings become sufficiently hardened they took the advantage of every breeze of North and Northern West Wind that blew until they all disappeared from our Vicinity. But Alas! They left the land a disolute wilderness. I believe there is not one single ear of Wheat or Barly in the whole Colony. There are a few small watery Potatoes, which may answer for seed. Hundred of Settlers are off to winter on the plains and others to fish at the Lakes, but god only know how the Multitude will be able to survive the Winter. The Crops of last season were good but the people were foolish and improvident. And the consequence is that large quantities of grain fell into the hands of persons who ground it up and sent it out of the colony into the interior to traffic with the Indians for Furs, and the growers of the grain run a risk of perishing by hunger during the Winter. And if they survive the winter they will, I fear, in hundreds of Cases be unprovided with seed in the spring. The Buffaloe hunt proved a total failure last summer. In fact these useful animals are nearly exterminated. And what the future has in store for the Indians and halfbreed of this land is a problem not easily solved. I had two notes last summer from Prof. Henry the last of them covering a receipt, this receipt I saved and returned but no order has yet come to hand. I am sorry to tell you that I had been confined to the House during the whole spring with Rheumatic pains in my left leg, knee and in both my hands so that I have not been able to do anything in collecting either birds or nests. I have just skinned a white owl. I have a large Grey owl skin and a Pelican skin. If you wish to have please let me know and I will forward them. I Remain My Dear Prof

Yours Most Sincerely

D. GUNN SERVT

P.S. Our summer was not very hot, a few days in July indicated 85 and as high on one occassion at night as 72, great quantities of rain fell during the month. On the morning of the third, a fearful Hurricane of wind and tempest of rain from the West passed over the Settlement knocking down buildings and killing a few persons. This month has been very frosty.D.G.

APPENDIX

Northern collectors often referred to zoological specimens by their scientific name. The Latinized binomes found in their letters were not always spelled correctly, but they were always similar enough to nineteenth century nomenclature to render recognition easy. Below I have provided the common names for the species referred to by scientific nomenclature in the letters written by northern correspondents. Although many of the names used by the nineteenth century scientific community have since been revised, with some of the species referred to in the foregoing letters having undergone several changes in the last one hundred years, no attempt has been made to rectify what must seem to be an anachronous presentation of mammalian, avian, ichthyological, and entomological taxonomy. The purpose of this Appendix is to clarify, as unobtrusively as possible, terminology that is uninteresting and generally inaccessible, except to scholars of taxonomic history.

Several sources have been used to construct this Appendix: S.F. Baird, J. Cassin and G. Lawrence, *The Birds of North America, The Descriptions of Species Based Chiefly on the Collections in the Museum of the Smithsonian Institution*, Natural Sciences in America, ed. Keir B. Sterling (Philadelphia: J.B. Lippencott & Co., 1860; reprint ed., New York: Arno Press, 1974); S.F. Baird, *The Mammals of North America, The Descriptions of Species Based Chiefly on the Collections in the Museum of the Smithsonian Institution*, Natural Sciences in America, ed. Keir B. Sterling (Philadelphia: J.B. Lippencott & Co., 1859; reprint ed., New York: Arno Press, 1974); C. Stuart Houston, *Arctic Ordeal, The Journal of John Richardson, 1820-1822*, with a Foreword by W. Gilles Ross (Kingston and Montreal: McGill-Queen's University Press, 1984); Samuel Eddy and James C. Underhill, *Northern Fishes*, With Special Reference to the *Upper Mississippi Valley*, Third ed. (Minneapolis: University of Minnesota Press, 1974); Charles T. Brues and A.L. Melander, *Key to the Families of North American Insects, An Introduction to the Classification of Insects* (Boston, Mass. and Pullman, Wash.: The Authors, 1915); and Ralph B. Swain, *The Insect Guide, Orders and Major Families of North American Insects*, Bureau of Entomology and Plant Quarantine, U.S. Department of Agriculture (Garden City, New York: Doubleday & Co., Inc., 1948).

BIRDS

Aegialitis (Aegialeus) semipalmatus	Semipalmated Plover
Aegiothus linaria	Lesser Red Poll
Agelaius gubernator	Red-shouldered Blackbird
Agelaius phoeniceus	Swamp Blackbird
Ampelis garrulus	Wax wing
Anas boschas	Mallard
Anser (Anser) gambelii	White-fronted Goose
Anser (Chen) hyperboreus	Snow Goose
Aquila canadensis	Golden Eagle; Ring-tailed Eagle
Archibuteo lagopus	Rough-legged Hawk
Archibuteo sancti-johannis	Black Hawk
Aythya vallisneria	Canvas-backed Duck
Bernicla (Bernicla) brenta	Brant
Bernicla (Leucoblepharon) canadensis	Canada Goose
Bernicla (Leucoblepharon) hutchinsii	Hutchin's Goose
Bernicla (Leucoblepharon) leucopareia	White-cheeked Goose
Bernicla (Leucopareia) leucopsis	Barnacle Goose
Bernicla (Bernical) nigricans	Black Brant
Bonasa umbellus	Ruffed Grouse
Brachyotus cassinii	Short-eared Owl
Bucephala albeola	Butter Ball
Bucephala americana	Golden Eye
Bucephala islandica	Barrow's Golden Eye
Buteo (Poecilopternis) borealis	Red-tailed Hawk
Calidris arenaria	Sanderling
Ceryle (Megaceryle) alcyon	Belted King-fisher
Charadrius virginicus	Golden Plover
Chroicocephalus philadelphia	Bonaparte's Gull
Circus hudsonius	Marsh Hawk
Colaptes auratus	Flicker
Collyrio borealis	Great Northern Shrike
Colymbus arcticus	Black-throated Diver

Colymbus septentrionalis	Red-throated Diver
Colymbus torquatus	Loon; Northern Diver
Corvus americanus	Common Crow
Corvus carnivorus	American Raven
Cotyle riparia	Bank Swallow
Cygnus americanus	American Swan
Cygnus buccinator	Trumpeter Swan
Dafila acuta	Sprig-tail; Pin-tail Duck
Dendroica aestiva	Yellow Warbler
Dendroica canadensis	Black-throated Blue Warbler
Dendroica coronata	Yellow Rump
Dendroica striata	Black Poll Warbler
Empidonax minimus	Least Flycatcher
Empidonax pusillus	Fly-catcher
Ereunetes petrificatus	Semipalmated Sandpiper
Erismatura rubida	Ruddy Duck
Falco (Hierofalco) candicans	Jer Falcon
Falco (Hierofalco) islandicus	Jer Falcon
Falco (Falco) nigriceps	Hawk
Falco (Tinnunculus) sparverius	Sparrow Hawk
Fulix affinis	Little Black-headed Duck
Fulix marila	Big Black-headed Duck
Gallinago wilsonii	English Snipe
Gambetta flavipes	Yellow Legs
Graculus dilophus	Double-crested Cormorant
Grus americanus	Whooping Crane
Grus canadensis	Sand-hill Crane
Haliaetus leucocephalus	Bald Eagle
Harelda glacialis	Duck
Lagopus albus	White Ptarmigan
Lagopus americanus	American Ptarmigan
Lagopus leucurus	White-tailed Ptarmigan
Lagopus rupestris	Ptarmigan
Larus glaucescens	The Glaucous-winged Gull

Larus glaucus	The Glaucous Gull
Larus leucopterus	The White-winged Gull
Macrorhamphus griseus	Red-breasted Snipe
Macrorhamphus scolopaceus	Snipe
Mareca americana	Baldpate; American Widgeon
Melanetta velvetina	Velvet Duck
Mergus americanus	Sheldrake
Mergus serrator	Red-breasted Merganser
Micropalama himantopus	Stilt Sandpiper
Nettion carolinensis	Green-winged Teal
Numenius (Phaeopus) hudsonicus	Hudsonian Curlew
Nyctale albifrons	Kirtland's Owl
Nyctale nivea	Snowy Owl
Nyctea richardsoni	Sparrow Owl
Parus hudsonicus	Titmouse
Passerculus anthinus	Sparrow
Passerculus sandwichensis	Sparrow
Passerculus savanna	Savannah Sparrow
Passerella iliaca	Fox-colored Sparrow
Pedioecetes phasianellus	Sharp-tailed Grouse
Pelionetta perspiicllata	Surf Duck
Phalaropus hyperboreus	Northern Phalarope
Pica hudsonica	Magpie
Picoides arcticus	Black-backed three-toed Woodpecker
Picoides dorsalis	Striped three-toed Woodpecker
Picoides hirsutus	Banded three-toed Woodpecker
Picus (Phrenopicus) borealis	Red-cockaded Woodpecker
Picus (Trichopicus) pubescens	Downy Woodpecker
Picus (Trichopicus) villosus	Hairy Woodpecker
Pinicola canadensis	Pine Grosbeak
Plectrophanes (Centropanes) pictus	Smith's Bunting

Plectrophanes (Centrophanes) lapponicus	Lapland Longspur
Plectrophanes (Plectrophanes) nivalis	Snow Bunting
Podiceps auritus	Eared Grebe
Ptychorhamphus aleuticus	Brandt
Rissa septentrionalis	The North Pacific Kittiwake
Scolecophagus ferrugineus	Rusty Blackbird
Scops asio	Mottled Owl
Setophaga ruticilla	Redstart
Somateria mollissima	Eider Duck
Somateria spectabilis	King Eider
Somateria V. nigra	Pacific Eider
Spatula clypeata	Shoveller Duck
Spizella monticola	Tree Sparrow
Spizella pallida	Clay-colored Bunting
Spizella pusilla	Field Sparrow
Squatarola helvetica	Black-bellied Plover
Stercorarius cepphus	Buffon's Skua
Stercorarius parasiticus	The Arctic Skua
Sterna caspia	The Caspian Tern
Sterna macroura	The Arctic Tern
Strepsilas interpres	Turnstone
Surnia ulula	Hawk Owl
Syrnium cinereum	Great Gray Owl
Tringa (Actodromas) wilsonii	Least Sandpiper
Tringoides macularius	Spotted Sandpiper
Tryngites rufescens	Buff-breasted Sandpiper
Turdus (Turdus) aliciae	Gray-cheeked Thrush
Turdus (Turdus) fuscescens	Wilson's Thrush
Turdus (Turdus) pallasi	Hermit Thrush
Xema sabinii	The Fork-tailed Gull
Zonotrichia albicollis	White-throated Sparrow
Zonotrichia coronata	Golden-crowned Sparrow
Zonotrichia gambelii	Sparrow

* * * * * * * * * *

MAMMALS

Arctomys monax	Woodchuck
Arctomys pruinosus	Hoary Marmot
Arvicola gapperi	Red-Backed Mouse
Arvicola richardsonii	British Rocky Mountain Marmot
Arvicola riparia	Bank Mouse
Arvicola xanthognathus	Hudson's Bay Marmot
Cervus canadensis	American Elk
Condylura cristata	Star-nosed Mole
Gulo luscus	Wolverine
Hesperomys myoides	Hamster Mouse
Lepus americanus	Northern Hare; White Rabbit
Lutra canadensis	American Otter
Myodes	Lemming
Ovibos moschatus	Musk Ox
Putorius longicauda	Long-Tailed Weasel
Putorius noveboracensis	White Weasel; Ermine
Putorius richardsonii	Little Ermine
Rangifer caribou	Woodland Caribou
Rangifer groenlandicus	Barren Ground Caribou
Sorex palustris	Marsh Shrew
Sorex richardsonii	Richardson's Shrew
Spermophilus franklini	Gray Gopher
Spermophilus parryi	Parry's Marmot
Ursus maritimus	White or Polar Bear
Vulpes fulvus var. fulvus	Red Fox

* * * * * * * * * *

FISH

Coregonus clupeaformis	Lake Whitefish
Salvelinus alpinus	Arctic Char
Salvelinus namaycush	Lake Trout
Salvelinus siscowet	Lake Trout
Stenodus leucichthys	Inconnu
Stenodus mackenzie	Inconnu or Mackenzie River Salmon

* * * * * * * * * *

INSECTS

Boreidae	Scorpion Flies
Braconidae	Ichneumon Flies
Brenthidae	Brentid Beetles
Chalcididae	Chalcid Wasps
Cimbicidae	Willow sawfly
Cossidae	Carpenter Moths
Pteronarcys regalis	Stoneflies
Syrphidae	Syrphid Flies
Tenthredinidae	Sawflies

INDEX